An Advanced Treatise on Fixture Design and Planning

SERIES ON MANUFACTURING SYSTEMS AND TECHNOLOGY

Published

Vol. 1: An Advanced Treatise on Fixture Design and Planning
Andrew Y. C. Nee, Z. J. Tao & A. Senthil Kumar

Series on Manufacturing Systems and Technology – Vol. 1

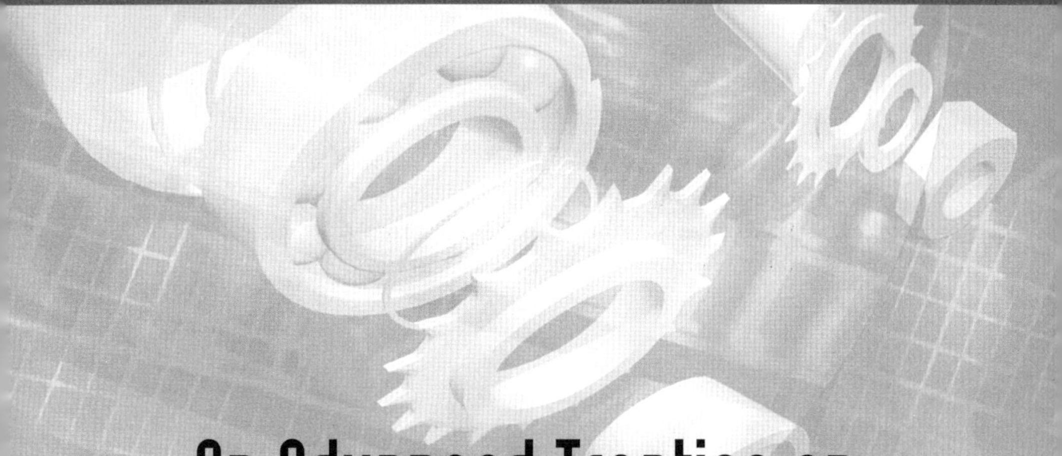

An Advanced Treatise on Fixture Design and Planning

Andrew Y. C. Nee
National University of Singapore

Z. J. Tao
Nanyang Polytechnic

A. Senthil Kumar
National University of Singapore

World Scientific

NEW JERSEY • LONDON • SINGAPORE • BEIJING • SHANGHAI • HONG KONG • TAIPEI • CHENNAI

Published by

World Scientific Publishing Co. Pte. Ltd.
5 Toh Tuck Link, Singapore 596224
USA office: 27 Warren Street, Suite 401–402, Hackensack, NJ 07601
UK office: 57 Shelton Street, Covent Garden, London WC2H 9HE

British Library Cataloguing-in-Publication Data
A catalogue record for this book is available from the British Library.

AN ADVANCED TREATISE ON FIXTURE DESIGN AND PLANNING

ISBN 981-256-059-9

Editor: Tjan Kwang Wei

Printed in Singapore by World Scientific Printers (S) Pte Ltd

A.Y.C. Nee	To my wife Siew Kheng and daughters Joanne, Julianne, Suzanne for their patience and understanding
Z.J. Tao	To my wife Hongbing for her patience and support
A. Senthil kumar	To my parents Prof Anantharajan and Ms Vijayalakshmi, and my wife Amutha and son Saravan for their help and understanding

Preface

The primary goal of this book is to cover, as much as possible, the state-of-the-art development in the domain of computer-aided fixture design and planning. The text concentrates on key issues central to the development of computer-integrated manufacturing, such as fixture design automation, fixture clamping layout synthesis, clamping intensity optimisation, workpiece-fixture interaction, intelligent fixtures which are integrated with processing equipment or machine tools, Internet-enabled fixture design and modular fixture database management.

This book is intended to be a reference text for academics, manufacturing and industrial engineers. It may also be used as a text for engineering graduate students in the discipline.

The organisation of the book is arranged in a topical manner. It begins with a concise presentation of the generic principles of fixture design in Chapter 1. Chapter 2 discusses the concept of integrated computer-aided fixture design system. It presents a 3D CAD-based system which is able to produce automated, interference-free fixture design and assembly solutions. Chapters 3 and 4 cover the framework and methodology to determine a viable clamping layout in terms of optimal clamping points, positive clamping actuation sequence, and minimal sustainable clamping intensities.

The dynamic interaction between a workpiece and its fixture plays a crucial role in fashioning the finished accuracy of a part. Experimental investigation presented in Chapter 5 provides an insight into the workpiece stability and the inherent dynamic nature of the workpiece-fixture system. The book then details the architecture of an intelligent fixturing system in Chapter 6, which is capable of tool path compensation and on-the-fly moderation of fixturing parameters so as to minimize workpiece displacement and distortion.

Chapter 7 presents a fixture element database management system which is essential to the automation of fixture design process. It is followed by a description of an Internet-based interactive fixture design system in Chapter 8. This system promotes the concept of collaborative design in a distributed manufacturing environment.

Acknowledgements

During the period of preparing for the book, we have received excellent support from our colleagues. We would like to acknowledge the valuable contributions by Dr J.Y.H. Fuh, Dr M.A. Mannan and Dr Y.S. Wong at the National University of Singapore. Our graduate students have also contributed to this book with their research work, particularly: S. Senthilkumar, J.R. Dai, L. Lin, X.W. Gui, J. P. Sollie, Y.F. Wang, T.K. Xiao and Long Qin, and in specific chapters: Chapter 2 – T.S. Kow; Chapter 7 – Y.M. Yip; and Chapter 8 – S.H Bok, Ratnapu Kiran Kumar, Sandeep Kumar Arya, Y.J. Ng and Fathianathan Mervyn. We would also like to thank K.W. Tjan at World Scientific for his support in the production of this book.

Last but not least, we would like to dedicate the book to three distinguished academicians who have contributed indirectly to this book, they are the late

Dr Ken Whybrew
Professor Amitabha Bhattacharyya
Professor Inyong Ham

Dr Ken Whybrew was the co-author of a first book "Advanced Fixture Design for FMS" written together with AYCN and ASK. Prof Bhattacharyya provided an unpublished manuscript written with Prof Ham to AYCN many years ago and some materials of this manuscript are presented in Chapter 1.

A.Y.C. Nee
National University of Singapore

Z.J. Tao
Nanyang Polytechnic, Singapore

A. Senthil kumar
National University of Singapore

Table of Contents

Chapter 3 Fixture Clamping Layout Synthesis

Chapter 4 Optimisation of Dynamic Clamping Forces for a Fixture

Chapter 5 Workpiece-Fixture Interaction

Chapter 8 An Internet-Enabled Smart Interactive Fixture Design System

Chapter 1

Introduction to Fixture Design

1.1 Introduction

A fixture is a device for locating, holding and supporting a workpiece during a manufacturing operation. Fixtures are essential elements of production processes as they are required in most of the automated manufacturing, inspection, and assembly operations.

Fixtures must correctly locate a workpiece in a given orientation with respect to a cutting tool or measuring device, or with respect to another component, as for instance in assembly or welding. Such location must be invariant in the sense that the devices must clamp and secure the workpiece in that location for the particular processing operation.

There are many standard workholding devices such as jaw chucks, machine vises, drill chucks, collets, etc. which are widely used in workshops and are usually kept in stock for general applications.

Fixtures are normally designed for a definite operation to process a specific workpiece and are designed and manufactured individually. Jigs are similar to fixtures, but they not only locate and hold the part but also guide the cutting tools in drilling and boring operations. These workholding devices are collectively known as *jigs and fixtures*. Figure 1.1 shows an example of a fixture commonly used on a horizontal CNC milling machine.

1.1.1 Elements of Fixtures

Generally, all fixtures consist of the following elements:

- Locators

A locator is usually a fixed component of a fixture. It is used to establish and maintain the position of a part in the fixture by constraining the movement of the part. For workpieces of greater variability in shapes and surface conditions, a locator can also be adjustable.

- Clamps

A clamp is a force-actuating mechanism of a fixture. The forces exerted by the clamps hold a part securely in the fixture against all other external forces.

- Supports

A support is a fixed or adjustable element of a fixture. When severe part displacement/deflection is expected under the action of imposed clamping and processing forces, supports are added and placed below the workpiece so as to prevent or constrain deformation. Supports in excess of what is required for the determination of the location of the part should be compatible with the locators and clamps.

- Fixture Body

Fixture body, or tool body, is the major structural element of a fixture. It maintains the spatial relationship between the fixturing elements mentioned above, *viz.*, locators, clamps, supports, and the machine tool on which the part is to be processed.

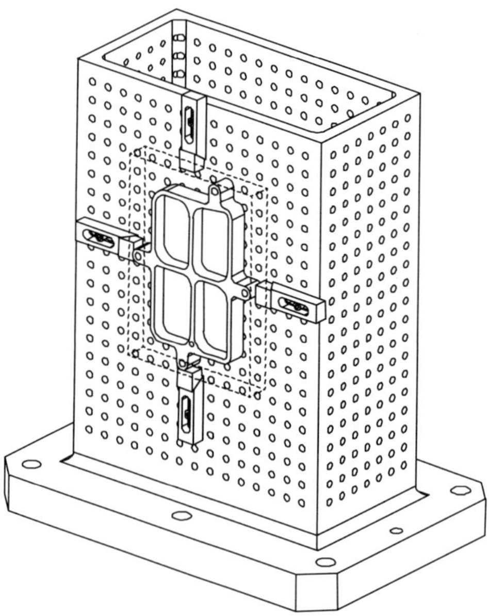

Figure 1.1 A typical 'tombstone' fixture for a horizontal CNC machine

1.1.2 Importance of Fixtures in Manufacturing

Modern manufacturing aims at achieving high productivity to reduce unit cost. This necessitates workholding devices to be efficient, *i.e.* to increase the rate of loading and unloading to speed up the manufacturing cycle time.

If t is the total time in seconds or minutes required for producing a part, then $Q = \frac{1}{t}$ is the number of pieces produced in unit time, or the production rate.

Considering the fact that the total manufacturing time is usually composed of:

$$t = t_m + t_h$$

where t_m is the actual machining time and t_h is the setting up and handling time, hence, the production rate is given by:

$$Q = \frac{1}{t_m + t_h} \text{ piece per unit time} \qquad (1.1)$$

Supposing Q_t is the ideal production rate whereby there is no handling time loss for a given machining operation, hence we have:

$$Q_t = \frac{1}{t_m}$$

Now,

$$Q = \frac{1}{\frac{1}{Q_t} + t_h} = \frac{1}{1 + \left(\frac{t_h}{t_m}\right)} Q_t = \lambda Q_t \qquad (1.2)$$

This factor $\lambda = \frac{1}{1 + \left(\frac{t_h}{t_m}\right)}$ can be termed as production efficiency.

The variation of λ with respect to Q_t is shown in Figure 1.2 for the various values of t_h. For an operation with a value of $t_m = t_h$, λ is 0.5 whereas, if $t_h = 2\, t_m$, λ is 0.33 and the production rate is reduced. Figure 1.3 shows how t_m and t_h affect production rate. It is clear from Figures 1.2 and 1.3 that

(a) For a given t_m, reduction of t_h increases Q,
(b) For a given t_h, reduction of t_m enhances Q.

The use of fixtures has twofold benefits. It eliminates individual marking, positioning and frequent checking before machining operation starts, thereby resulting in considerable saving in set-up time. In addition, the usage of work-holding devices saves operator labour through simplifying locating and clamping tasks and makes possible the replacement of skilled workforce with semi-skilled labour, hence effecting substantial saving in labour cost which also translates into enhanced production rate.

Furthermore, the use of well-structured fixtures with higher locating and clamping rigidity would allow for increase in cutting speeds and feeds, thereby reducing t_m, hence improving production rate.

Besides improving the productivity in terms of the rate of production, there are also other benefits accrued through the use of fixtures. They are:

(a) Increases machining accuracy because of precise location with fixtures,
(b) Decreases expenditure on quality control of machined parts as fixtures facilitate uniform quality in manufacturing,
(c) Widens the technology capacity of machine tools and increases the versatility of machining operations to be performed,
(d) Either fully or partly automates the machine tool.

1.1.3 General Requirements of a Fixture

In order to maintain the workpiece stability during a machining process, an operational fixture has to satisfy several requirements to fully perform its functions as a workholding device. The following constraints must be observed while designing a viable fixture:

- Deterministic location

A workpiece is said to be kinematically restrained when it cannot move without losing contact with at least one locator. The workpiece is constrained by a set of appropriately placed locators so that it is presentable for the machining operation. Locating errors due to locators and locating surfaces of the workpiece should be minimised so as to accurately and uniquely position the workpiece within the machine coordinate frame.

- Total constraint

A workpiece should be fully constrained at all times to prevent any movement. Clamps should provide locking forces to hold the workpiece in place once it is located. A totally restrained part should be able to remain in static equilibrium to withstand all possible processing forces or disturbance. A necessary and sufficient condition to warrant workpiece stability is to satisfy the condition of force closure.

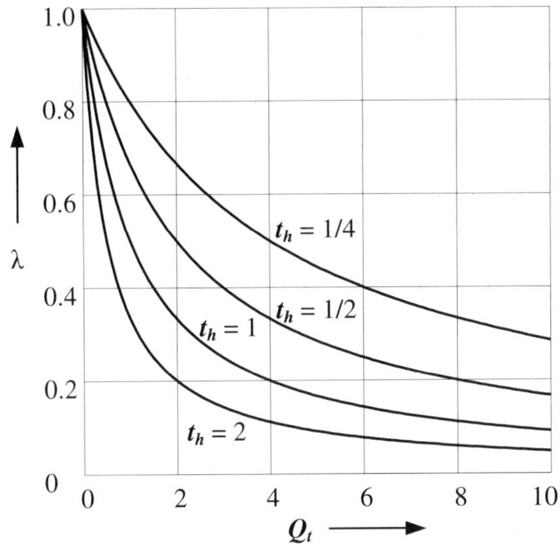

Figure 1.2 Effect of setting and handling time (t_h) on production efficiency (λ).

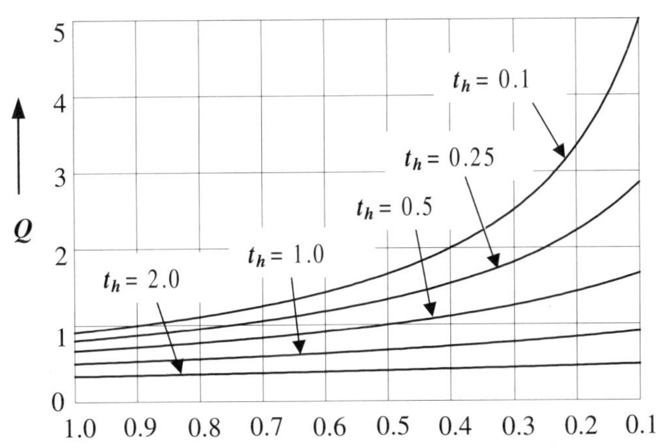

Figure 1.3 Effect of handling and machining times (t_h & t_m) on production rate (Q).

- Contained deflection

Workpiece deformation is unavoidable due to its elastic/plastic nature, and the external forces impacted by the clamping actuation and machining operations. Deformation has to be limited to an acceptable magnitude in order to achieve the tolerance specifications.

- Geometric constraint

Geometric constraint guarantees that all fixturing elements have an access to the datum surface. They also assure that the fixture components do not interfere with cutting tools during a machining operation.

In addition to these requirements, a fixture design should have desirable characteristics such as quick loading and unloading, minimum number of components, accessibility, design for multiple cutting operations, portability, low cost, etc.

1.1.4 Fixture Design Fundamentals

Fixture design consists of a number of distinct activities: fixture planning, fixture layout design, fixture element design, tool body design, etc. They are listed in Figure 1.4 in their natural sequence, although they may be developed in parallel and not necessarily as a series of isolated activities in actual execution.

Fixture design deals with the establishment of the basic fixture concepts:

- Fixture layout is an embodiment of the concepts in the form of a spatial configuration of the fixture,
- Fixture element design is concerned with the concrete details of the locators, clamps and supports,
- Tool body design produces a structure combining the fixture elements in the desired spatial relationship with the machine tool.

1.1.4.1 Fixture Design

Fixture planning is to conceptualise a basic fixture configuration through analysing all the available information regarding the material and geometry of the workpiece, operations required, processing equipment for the operations, and the operator. The following outputs are included in the fixture plan:

- Fixture type and complexity
- Number of workpieces per fixture

- Orientation of workpiece within fixture
- Locating datum faces
- Clamping surfaces
- Support surfaces, if any

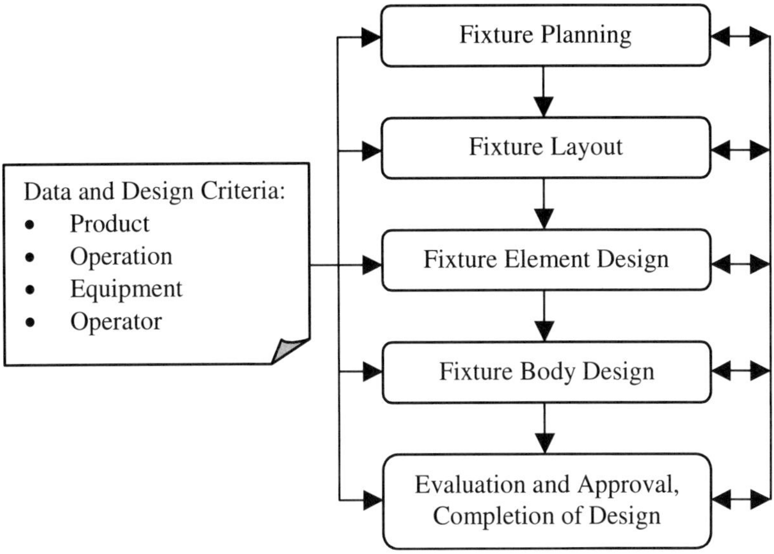

Figure 1.4 Various aspects of fixture design

Generation of fixture layout is to represent the fixture concepts in a physical form. The following outputs are included in the fixture layout:

- Positions of locators
- Positions of clamps
- Positions of supports, if any
- Type of locators
- Type of clamps
- Type of supports
- Clamping forces and sequence

Fixture element design is either to detail the design drawings committed on paper or to create the solid models in a CAD system of the practical embodiment of the conceptual locators, clamps and supports. It is possible to use standard designs or proprietary components. The following outputs are included in the fixture element design:

- Detailed design of locators

- Detailed design of clamps
- Detailed design of supports, if any

Tool body design is to produce a rigid structure carrying all the individual fixture elements in their proper places.

1.1.4.2 Fixture Design Criteria

The following design criteria must be observed during the procedure of fixture design:

- Design specifications
- Factory standards
- Ease of use and safety
- Economy

1.1.4.3 Fixture Design Procedure

In the design of a fixture, a definite sequence of design stages is involved. They can be grouped into three broad stages of design development.

Stage One deals with information gathering and analysis. These include product analysis such as the study of design specifications, process planning, examining the processing equipment and considering operator safety and ease of use. In this stage, all the critical dimensions and feasible datum areas are examined in detail.

Stage Two involves the consideration of clamping and locating schemes. A clamping scheme is devised in such a way that it will not interfere with the tools or cutters and are fully compatible with proposed locating surfaces or areas. The locating scheme, using standard elements such as pins, pads, etc. is designed to be consistent with clamping and tool-guiding arrangements.

Stage Three is the design of the structure of the fixture body frame. This is usually built around the workpiece as a single element which links all the other elements used for locating, clamping tool-guiding, etc. into an integral frame work.

The above procedures are quite general and can be modified depending on the relative importance of the various elements in providing for the required accuracy of the workpiece to be located and secured into the fixturing device. With the popular adaptation of modular fixturing elements, the fixture body frame is usually a standard block with fixed arrays of locating and fixing holes or slots. It becomes a matter of selecting the most suitable body frame to accommodate the various elements, provide good support of the workpiece and access to cutters and tools. Figure 1.1 shows a 'tombstone' body frame commonly used in horizontal CNC milling machines.

1.2 Locating Principles

1.2.1 Introduction

One of the principal purposes of a machining fixture is to locate the workpiece surfaces for performing a machining operation. This is usually done with respect to a number of factors to be considered such as the reference datum, supporting surfaces, features that are likely to obstruct the tool movement or access direction, etc. In general, the following surfaces should be distinguished:

- Active surfaces

These are surfaces to be machined, *i.e.* surfaces which are subjected to the action of cutting tools.

- Supporting and locating surfaces

These are surfaces by means of which the workpiece is to be located with respect to set-to-size cutting tools.

- Clamping surfaces

Clamping surfaces are subjected to the clamping forces for obtaining invariant location. Clamping surfaces are usually not finish-machined surfaces as clamping marks could damage the finish.

- Datum surfaces

Datum surfaces are reference surfaces where the dimensions are to be maintained and measured.

- Free surfaces

Free surfaces are surfaces not involved in the set-up for the particular machining operation.

1.2.2 Restrictions on the Degrees of Freedom of a Workpiece

A workpiece, just like any free solid body, has six degrees of freedom (some researchers have referred this to the twelve degrees of freedom by considering the +/- movements in each category):

- Three rectilinear displacements along the mutually orthogonal co-ordinate axes
- Three angular displacements with respect to the same axes.

During a set-up, it is necessary to restrict certain degrees of freedom so as to locate and orient the active surfaces with respect to the cutting tools. Since supporting or restricting surfaces may vary from the true geometrical shape, especially on rough-machined surfaces or cast blanks, it is desirable that the workpiece be located with respect to the point supports.

Locating using point supports in the form of hemi-spherical rest buttons would considerably reduce the influence of geometrical variations of locating surfaces on the locating accuracy. For prismatic parts, the general principle of 3-2-1 location is most commonly employed. For achieving greatest stability, the first three points of location on the primary surface should be as far apart as possible, or the area enclosed by the three points as large as possible.

For larger cast workpieces, the 4-2-1 locating principle is frequently used. Since this violates the locating constraints, one of the locating points would need to be an adjustable one. However, it is also a good practice for larger castings to be designed with accurate fixturing points. These points are subsequently removed after the first few surfaces have been machined.

For cylindrical workpieces, three-point location cannot be obtained because of the non-existence of plane surfaces, V-locators and close-fitting bushes are often used instead. For circular laminae, location can be achieved with the aid of a slot-support.

When a workpiece is required to be located with respect to an inside hole or bore, a plug is used for locating the workpiece. Locating from two holes typically uses a full and a diamond plug combination, with the latter inserted in the larger of the two holes.

The details of the principles of location can be easily found in general texts on fixture design (Henriksen, 1973) and will not be repeated here.

1.3 Clamping Principles

1.3.1 Introduction

In every machining operation, clamping of workpieces is an essential requirement. A clamp can be defined as a device for providing an invariant location with respect to an external loading system. In other words, the process of clamping induces a locking effect which, through frictional or some other forms of mechanism, provides a stability of location which cannot be changed until and unless external loading is able to overcome the locking effect. Hence, when a cutting force is producing a load or moment on the workpiece, it is necessary that a sufficient clamping force must be exerted to withstand such actions. The creation and retention of locking effect against external loads are the principal objectives of any locking devices. The generalised requirements of locking elements can be summarised as:

- To provide a suitable locking for achieving the stability of location,
- To produce sufficient frictional effects for the above purpose but without causing any undesirable effects to the workpiece such as distortion or surface damage.

It is also essential that the idle time involving loading, locking, unlocking and unloading of workpieces should be minimised as much as possible to reduce the overall set-up and non-machining time. Certain additional requirements are therefore to be fulfilled with respect to clamping devices:

- The clamping devices must be easy to manipulate manually or otherwise,
- These devices must be quick-acting so as to reduce time for setting the clamping and simultaneous locating,
- They must be low-cost so that their application in small lot sizes is economical.

1.3.2 Basic Principles of Clamping

1.3.2.1 Orientation of Locators vis-à-vis Clamping Force

It is necessary in all clamping devices that the clamping forces hold the workpiece in its located position and should not cause any positional displacement or excessive distortion under the action of the clamping forces.

Clamping forces should be directed towards supporting and locating elements on overhanging or thin sections of the workpiece. In addition, the force should be transmitted to the rigid sections of the body frame of the fixture.

Cylindrical workpieces located in V-blocks can be clamped using another V-block, making a 4-point clamping, or clamped in a 3-jaw chuck, in a 3-point clamping configuration. The latter is usually more common, especially in turning operations.

1.3.2.2 Effect of External Forces on the Clamping Action

Clamping elements can be classified in accordance with their force-deflection characteristics. There are two broad sub-divisions, *viz.*:

- Type I: clamping elements in which the elastic deformation increases with clamping force, such as screws, levers, cams, etc.,

- Type II: clamping elements in which the clamping force assumes a constant value independent of the elastic deformation at the contact surfaces such as fixtures operated with hydraulic or pneumatic pressures.

Within the elastic region, clamping elements based on elastic deformation, *i.e.* Type I clamps, would exhibit a linearly increasing clamping force in proportion to the deformation of the clamping element, if the workpiece or the locator is assumed to be rigid. If the workpiece or locator deforms, it will cause a relaxation of the clamping element and the clamping force will decrease. A limiting case arises when the clamping is lost and the force becomes zero.

In Type II clamps, the clamping force remains constant at pre-set values and is independent of workpiece and locator deformation. This type of clamping device is therefore more reliable and would not relax over time.

1.3.2.3 Types of Clamps

Clamping elements may be either manually operated or actuated by pneumatic, hydraulic or a combination of other power facilities. They are also classified according to the mechanism by which a mechanical advantage is attained. The two basic classes include:

- Application of inclined plane theory, *i.e.* wedges, screws, cams, etc.,
- Application of lever principle, *i.e.* levers, toggles, etc.

Manual clamping of workpieces has the following disadvantages:

- Each workpiece is clamped with varying force,
- It is difficult to determine the required force for reliable clamping,
- Fatigue of operator due to manual clamping takes place,
- Time required to actuate manual clamping is longer compared to power-actuated clamping,
- Comparatively small amount of force is available without large force amplification devices.

Pneumatic and hydraulic clamping devices have eliminated most of the above disadvantages but at much higher cost as well as greater demand for space requirement and maintenance. Justification would be a balance between cost, efficiency, accuracy, operator safety and comfort. As will be explained in the subsequent sections of this book, clamping with such devices forms the basis of variable-force clamping, which is very useful in controlling the intensity of the clamping force during a machining operation and helps to reduce workpiece deformation.

For the design of non-standard clamping devices, the most comprehensive source of information can be found in a resource book, *Jigs and Fixtures: Non-Standard Clamping Devices* (Grant, 1967).

1.4 Automation in Fixture Design

1.4.1 The Need for More Flexible Fixtures

With the advent of CNC machining technology and the capability of multi-axis machines to perform several operations and reduce the number of set-ups, the fixture design task has been somewhat simplified in terms of the number of fixtures which would need to be designed. However, there is a need to address the faster response and shorter lead-time required in designing and constructing new fixtures. The rapid development and application of Flexible Manufacturing System (FMS) has added to the requirement for more flexible and cost-effective fixtures. Traditional fixtures (dedicated fixtures) which have been used for many years are not able to meet the requirements of modern manufacturing due to the lack of flexibility and low reusability. The replacement of dedicated fixtures by modular and flexible fixtures is eminent in automated manufacturing systems, due to much smaller batch sizes and shortened time-to-market requirement.

Modular fixtures are constructed from standard fixturing elements such as base-plates, locators, supports, clamps, etc. These elements can be assembled together without the need of additional machining operations and are designed for reuse after disassembly (Nee *et al.,* 1995). The main advantages of using modular fixtures are their flexibility and the reduction of time and cost required for the intended manufacturing operations. Automation in fixture design is largely based on the concept of modular fixtures, especially the grid-hole-based systems, due to the following characteristics:

- Predictable and finite number of locating and supporting positions which allow heuristic or mathematical search for the optimum positions,
- Ease in assembly and disassembly and the potential of automated assembly using robotic devices,
- Relative ease of applying design rules due to the finite number of element combinations.

A typical modular fixture constructed from the grid-hole-based system is shown in Figure 1.5.

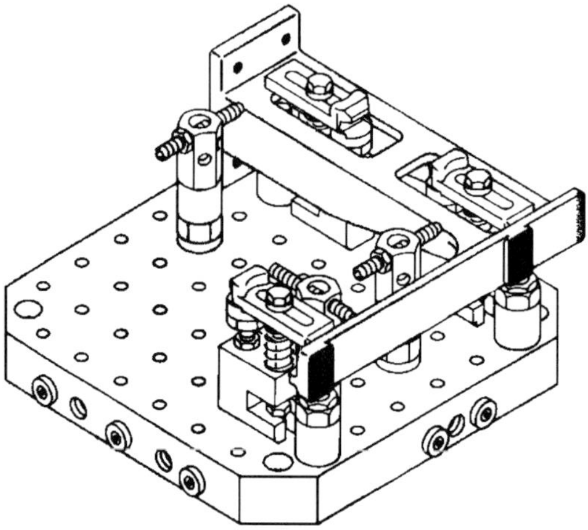

Figure 1.5 A fixture constructed from modular elements (Courtesy of IMAO Corp.)

1.4.2 Computer-aided Fixture Design Research

Fixture design research employing computer aids started in the late 1970s and early 1980s. In the initial years, interactive and semi-interactive fixture design techniques were built on top of commercial CAD systems and expert system tools. These approaches were mainly concerned with fixture configuration and there was little analysis on the other aspects such as workpiece-fixture-cutting tool interactions.

A comprehensive fixture research plan should involve the analysis at different computational levels, *viz.*, geometric, kinematic, force and deformation analyses. The following sections will present brief overviews of the research activities in each of the above-mentioned areas, followed by the need to design an intelligent fixture which can be integrated with the machine tool. This is also the aim of the main sections of the book.

1.4.2.1 Geometric Analysis

Geometric analysis is closely associated with fixture planning and spatial reasoning. It determines the selection of the type and number of fixturing elements, support and locating elements, the order of datum planes, etc. The analysis also includes the checking of interference between workpiece and fixturing elements, as well as cutting tools.

Most of the early fixture research involved geometric analysis and synthesis of fixture construction with relatively little attention to kinematic and deformation analysis.

Some of the earlier research works are briefly described. (Gandhi and Thompson, 1986) proposed a methodology for the automated design of modular fixtures. Some of the semi-automated systems were developed by (Jiang *et al.*, 1988), (Grippo *et al.*, 1987) and (Markus, 1988). (Nee and Senthil kumar, 1991) developed an automated fixture design system that integrates CAD with an expert system shell. Other systems using a 3-D solid modeller, an expert-system shell and a rule-based technique have been reported by (Lim and Knight, 1986), (Nnaji *et al.*, 1988), (Nnaji and Lyu, 1990), and (Ngoi *et al.*, 1990). These systems are capable of automatically generating partial or complete fixturing solutions for simple prismatic workpieces, based mostly on geometric analysis. Several of the above systems are based on modular fixtures.

1.4.2.2 Kinematic Analysis

Kinematic analysis is used to determine whether a fixture configuration is able to correctly locate and provide complete constraint to a workpiece.

Previous work on fixture design automation offers relatively little consideration in providing a comprehensive fixture-element database and effective assembly strategies for the generation and construction of modular fixtures. The assembly of modular fixtures is to configure the fixture elements such as locators, clamps and supports (in most cases, accessory elements are needed to generate fixture towers to fulfil the fixturing functions) on the base-plate according to a fixturing principle (e.g. 3-2-1 principle as outlined in Section 1.2.2). The determination of the locating, supporting and clamping points for the assembly of modular fixtures is a key issue in fixture design automation. (Trappey and Matrubhutam., 1993) proposed a heuristic approach using the projected envelope of a workpiece to determine the locating and clamping points. This method reduces the complexity of fixture configuration and could be feasible for non-prismatic parts, but the system does not consider the size of the modular fixture elements and interference between the elements, the machining envelope and the workpiece. (Chou, 1994) employed the geometric reasoning method to determine the locating and clamping points, but a detailed assembly method has not been proposed. (Whybrew and Ngoi, 1992) proposed a method using the spatial representation technique for automatic assembly of modular fixturing elements. The system uses a T-slot modular fixture system, however, the positions and types of locating, clamping and supporting elements are given as the inputs to the system. Treating a machined workpiece as a rigid body, (DeMeter, 1994a & 1994b) applied restraint analysis to a fixture with frictionless or frictional surface contacts and a linear model for predicting the impact of locator and clamp placement on workpiece displacement throughout machining operations.

In modular fixture design and assembly, particularly for a hole-matrix system, the exact points of locating, supporting and clamping should not be pre-determined as inputs before actual assembly. Since the fixture elements may not fit directly onto the tapped and location holes, it would be more difficult and sometimes not feasible to assemble the fixture elements properly. Hence, it is preferable to determine the configuration of the fixture from an assembly viewpoint.

1.4.2.3 Force Analysis

In a machining fixture, different forces are experienced, *viz.*, inertial, gravitational, machining and clamping forces. While the first three categories of forces are usually more predictable, clamping force can be rather subjective in terms of magnitude, point of application as well as sequence of application.

It has been widely accepted that a thorough analysis of all the forces involved in a fixture is a formidable task since it is an indeterminate problem with a large number of fixturing elements. When friction is taken into account, the problem becomes even more complex because both the magnitude and the direction of the static friction forces are unknown. (Lee and Haynes, 1987) reported that friction is the predominant mechanism for workpiece holding in most fixturing applications.

Even with the same fixturing forces, the force distribution in fixtures may vary significantly with different application sequence of clamping forces (Cogun, 1992). (Mittal *et al.*, 1991) and (Nnaji *et al.*, 1990) ignored the friction forces in their analysis. This simplification would lead to higher than necessary fixturing forces.

To analysis friction, (Lee and Cutkosky, 1991) constructed the limit surface in force/moment space as a convenient formalism to check the stability of the parts and to specify clamping forces. This method requires a search in the infinite clamping plane, and lacks theoretical sufficiency and is rather inefficient. When used for 3D analysis that is usually required for most fixturing configurations, this method is too complex and time-consuming to be applied. (Fuh and Nee, 1994) assumed the workpiece fixture as perfectly rigid bodies in frictional contact. (Gui *et al.*, 1996) employed linear springs to approximate the stiffness characteristics of contact between workpiece and fixture for minimising workpiece location error by optimising clamping force. (Jeng *at al.*, 1995) proposed a method to analyse the minimum clamping forces derived from the correlation between cutting force and clamping moment. This method increases the search efficiency by pruning inadequate search directions. Several stability criteria were set up after theoretical derivation, but these conclusions are correct for the force analysis with only one clamping plane. (Tao *et al.*, 1999a & 1999b) proposed a computational geometry approach to optimum clamping synthesis of machining fixtures. (Mannan and Sollie, 1997) designed a force control clamping system using feedback control and it provides an effective means for a variable force clamping system. (Li and Melkote, 2000) reported a minimum clamping force algorithm for machining fixtures.

1.4.2.4 Deformation Analysis

Due to the complexity of force interaction, workpiece deformation can be attributed to a combination of factors. Firstly, a workpiece would deform/displace under high cutting and clamping forces. Secondly, a workpiece could also deform/displace if the support and locating elements are not rigid enough to resist the above-mentioned forces. In the present analysis, it is assumed that workpiece deformation is largely due to the first cause mentioned above.

The most commonly used method in analysing workpiece deformation and fixturing forces is the finite-element method. (Lee and Haynes, 1987) and (Pham and Lazaro, 1990) studied the deformation of the workpiece, the clamping forces of the fixturing elements, the stress distribution and other characteristics of the fixturing system by modelling the workpiece as a deformable body. Although FEM can help understand the behaviour of the workpiece fixtured, it may not be a good choice for some applications due to its complexity.

1.4.3 Novel Clamping System Design

A good fixture design is critical to the quality of the finished workpiece in terms of dimensional accuracy, form precision and surface finish. One of the essential considerations in designing a good workable fixture is the generation of clamping configuration that includes the clamp placement, clamping sequence, and clamping intensities. Placing the clamps in wrong positions may disturb the equilibrium of the workpiece on the locators, resulting in the lost position of the part. Likewise, using an inadequate clamping intensity may give rise to slippage and/or lift-off of the workpiece during the machining process. On the other hand, an application of excessive clamping forces would result in excessive deflection and high contact deformation of the workpiece. In short, a poor clamping layout could cause the final accuracy of the workpiece to be out of the specified tolerances and bring about unnecessary rejects.

A less addressed research area is the performance of a fixture during machining in terms of its dynamic response and deformation. The issue is to guarantee machining accuracy through the proper control of workholding operation during machining. Therefore, a best approach to the fixturing problem is to integrate optimal fixture design with optimum fixturing execution in a unified approach.

This book devotes much effort on the development of an intelligent fixturing system (Nee *et al.,* 2000). This system provides sensory feedback and on-line fixturing control strategy to perform an optimal workholding operation. Being an important part of the "live" fixture, a novel dynamic clamping actuator capable of providing time-varying clamping intensities has been implemented. Comparative experiments are carried out to investigate the effects of the dynamic fixturing nature of the system on workpiece quality. Measured geometric errors are compared with and without using the dynamic clamping forces.

1.5 Summary

This chapter presents a brief introduction to the general and classic principles of fixture design. These principles are still very applicable and form the basis of good fixture design despite the rapid advancement of machine tools and manufacturing technology. As there are many classic texts dealing with the principles of location, clamping, and tolerance analysis of fixture design, these topics will not be covered in depth here and the readers are referred to such texts in the Bibliography.

With reduced time-to-market and smaller batch size, the use of dedicated fixtures has been steadily replaced with reconfigurable and reusable modular fixtures. Modular fixtures take a shorter time to configure and assemble, reduce inventory holding, and offer good repeatability and accuracy. They are invaluable in prototyping and proof-of-concept designs of new products. More importantly, they readily lend themselves to computerisation in design, analysis and assembly. Much of the automation in fixture design and analysis has been based on modular fixtures and this will continue to evolve in the future. Active research in this area has contributed much design knowledge and rules in fixture automation in the last 20 years. This chapter summarises some of the major contributions made by the fixture research community at the time of writing.

An ideal fixture should not only provide the machining repeatability and high productivity, it should also offer a solution which reduces workpiece distortion due to clamping and machining forces. The remaining sections of this book provide an in-depth analysis of this topic, which has not been addressed in depth previously.

1.6 Bibliography

Asada, H. and A.B. By. Kinematic Analysis of Workpart Fixturing for Flexible Assembly with Automatically Reconfigurable Fixtures. *IEEE Journal of Robotics and Automation*, 1(2), pp. 86-94. 1985.

Chou, Y.C. Geometric Reasoning for Layout Design of Machining Fixtures. *Int. J. Computer Integrated Manufacturing*, Vol.7, No.3, pp175-185. 1994.

Cogun, C. The Importance of the Application Sequence of Clamping Forces. *ASME Journal of Engineering for Industry*, Vol. 114, pp. 539-543. 1992.

DeMeter, E.C. Restraint Analysis of Fixtures which Rely on Surface Contact. *Journal of Engineering for Industry*, 116(2), pp. 207-215. 1994a.

DeMeter, E.C. The Min-Max Load Criteria as a Measure of Machining Fixture Performance. *Journal of Engineering for Industry*, 116(11), pp.500-507. 1994b.

Fuh, J.Y.H. and A.Y.C. Nee. Verification and Optimisation of Workholding Schemes for Fixture Design. *Journal of Design and Manufacturing*, 4, pp.307-318. 1994.

Gandhi, M.V. and B.S. Thompson. Automated Design of Modular Fixtures for Flexible Manufacturing Systems. *Journal of Manufacturing Systems*, 5(4), pp.243-252. 1986.

Grant, H.E. *Jigs and Fixtures: Non-Standard Clamping Devices*. New York: McGraw-Hill. 1967.

Grippo, P.M., M.V. Gandhi and, B.S. Thompson. The Computer Aided Design of Modular Fixturing Systems. *Int. J. Adv. Manuf. Tech.,* 2(2), pp.75-88. 1987.

Gui, X.W., J.Y.H. Fuh and A.Y.C. Nee. Modeling of Frictional Elastic Fixture-Workpiece System for Improving Location Accuracy. *IIE Transactions*, Vol.28, pp.821-827. 1996.

Henriksen, E.K. *Jig and Fixture Design Manual*. New York: Industrial Press. 1973.

IMAO. *IMAO Venlic Block Jig System (BJS) Fixturing Guide*. IMAO Corporation.

Institution of Production Engineers. *Data Memoranda on Jig and Fixture Design*. London: The Institutions of Production Engineers. 1973.

Jeng, S.L., L.G. Chen and W.H. Chieng. Analysis of Minimum Clamping Force. *Int. J. Mach. Tools Manufact.*, 35(9), pp.1213-1224.1995.

Jiang, W., S. Wang and Y. Cai. Computer-aided Group Fixture Design. *Annals of the CIRP,* 37(1), pp.145-148. 1988.

Lee, J.D. and L.S. Haynes. Finite-Element Analysis of Flexible Fixturing System. *Journal of Engineering for Industry*, 109(2), pp.134-139. 1987.

Lee, S.H. and M.R. Cutkosky. Fixturing Planning with Friction. *Journal of Engineering for Industry*, 113, pp.320-327. 1991.

Li, B. and S.N. Melkote. A Minimum Clamping Force Algorithm for Machining Fixtures. *Trans. of NAMRI/SME*, Volume XXVIII, pp.407-412. 2000.

Lim B.S. and J.A.G. Knight. Holdex - Holding Device Expert System. *Proc.1st Int. Conf. Applications of Artificial Intelligence in Engineering Problems,* pp.483-493. 1986.

Mannan, M.A. and J.P. Sollie. A Force-Controlled Clamping Element for Intelligent Fixturing. *Annals of the CIRP*, Vol. 46, pp.265-268. 1997.

Markus, A. Strategies for the Automated Generation of Modular Fixtures. *ASME Proc. Manufacturing International '88 Symposium Manufacturing Systems: Design, Integration and Control,* Vol.3, pp.97-103. 1988.

Mittal, R.O., P.H. Cohen and B.J. Gilmore. Dynamic Modelling of the Fixture-Workpiece System. *Robotics and Computer-integrated Manufacturing*, 8(4), pp.201-217. 1991.

Nee, A.Y.C. and A. Senthil kumar. A Framework for an Object/rule-based Automated Fixture Design System. *Annals of the CIRP*, 40(1), pp.147-151. 1991.

Nee, A.Y.C., A. Senthil kumar and Z.J. Tao. An Intelligent Fixture with a Dynamic Clamping Scheme. *Proceedings of the Institution of Mechanical Engineers*, 214(Part B), pp.183-196. 2000.

Nee, A.Y.C., A. Senthil kumar, S. Prombanpong and K.Y. Puah. A Feature-Based Classification Scheme for Fixtures. *Annals of the CIRP*, Vol.41, No.1, pp.189-192. 1992.

Nee, A.Y.C., K. Whybrew and A. Senthil kumar. *Advanced Fixture Design for FMS*. London: Springer-Verlag. 1995.

Ngoi, B.K.A., K. Whybrew and R.J. Astley. Computer Aided Assembly of Modular Fixturing Systems. *Proc. Asia-Pacific Industrial Automation '90 Conf. Automation Singapore,* pp.593-607. 1990.

Nnaji, B.O. and P. Lyu. Rules for an Expert Fixturing System on a CAD Screen Using Flexible Fixture. *Journal of Intelligent. Manufacturing,* Vol.1, pp.31-48. 1990.

Nnaji, B.O., S. Alladin and P. Lyu. A Framework for a Rule-Based Expert Fixturing System for Face Milling Planar Surfaces on a CAD System Using Flexible Fixtures. *Journal of Manufacturing. Systems,* 7(3), pp.193-207. 1988.

Pham, D.T. and A. de Sam Lazaro. Finite Element Study of a Workpiece in a Machining Fixture. *Mathematical and Computer Modelling*, Vol.14, pp.1024-1028. 1990.

Reulaux, F. *The Kinematics of Machinery*. New York: MacMillan. 1876. Reprint, New York: Dover. 1963.

Rong, Y. and Y. Zhu. *Computer-aided Fixture Design*. Marcel Dekker. 1999.

Sakurai, H. *Automatic Set-up Planning and Fixture Design for Machining*. PhD thesis, Massachusetts Institute of Technology. 1990.

Senthil kumar, A., A.Y.C. Nee and S. Prombanpong. Expert Fixture Design System for an Automated Manufacturing Environment. *Computer-Aided Design,* 23(6), pp.316-326. 1992.

Tao, Z.J., A. Senthil kumar and A.Y.C. Nee. A Computational Geometry Approach to Optimum Clamping Synthesis of Machining Fixtures. *International Journal of Production Research*, 37(15), pp.3495-3517. 1999b.

Tao, Z.J., A. Senthil kumar and A.Y.C. Nee. Automatic Generation of Dynamic Clamping Forces for Machining Fixtures. *International Journal of Production Research*, 37(12), pp.2755-2776. 1999a.

Trappey, A.J.C. and C.R. Liu. An Automatic Workholding Verification System. *Robotics and Computer-integrated Manufacturing*, 9(4/5), pp.321-326. 1992.

Trappey, A.J.C. and S. Matrubhutam. Fixture Configuration Using Projective Geometry. *Journal of Manufacturing Systems*, 12(6), pp.486-495. 1993.

Whybrew, K. and B.K.A. Ngoi. Computer Aided Design of Modular Fixture Assembly. *International Journal of Advanced Manufacturing Technology*, 7(5), pp. 267-276. 1992.

Chapter 2

Computer Aided Conceptual Fixture Design

2.1 Introduction

Traditionally, fixture design has been carried out manually and extensive heuristics knowledge from the designer is needed. This is time-consuming and reduces the productivity. In view of the trend towards advanced manufacturing techniques (AMT), requirement for an integrated and efficient fixture design system is becoming increasingly important. With the aid of such system, the process of fixture design can be automated and integrated with other manufacturing modules, which will lead to higher productivity and shorter manufacturing lead time, in accordance with the aims of automated manufacturing

In general, existing fixture design systems can be categorized into three different categories based on their degree of automation, namely, interactive, semi-automated and automated system (Nee *et al.*, 1995). An interactive fixture design system facilitates the design process by having informative user-interfaces for a user, based on his knowledge, to decide on the fixturing points and elements to configure a fixture set-up. Semi-automated fixture design system is improved from interactive systems where full knowledge from a designer is not required as either the fixturing points or fixture elements are determined automatically by the system. An automated fixture design system is one that automates the selection of fixturing points and elements by the use of certain design techniques such as rule-based design, case-based design, or other AI tools. In the process of design automation, CAD technology plays a vital role in modelling the fixtures geometrically and rendering the design solution graphically.

Several researchers have put in efforts to improve the fixture design process. Initial research works were mainly focused on the development of interactive computer-aided fixture design systems (Grippo *et al.*, 1987; Bai and Rong, 1995). Various works on design strategies and methodologies, such as the spatial relationships (Gandhi and Thomson, 1986), combination and conflict-resolution strategies (Markus, 1988), knowledge representation scheme (Darvishi and Gill, 1988), and kinematics constraint analysis (Bausch and Youcef-Toumi, 1990), have been applied to the design of fixtures. Apart from analytical approach, feature-based methodology for fixture design was developed to select fixturing faces and elements efficiently (Dong *et al.*, 1991). The feature-based representation schemes were developed as the basis for a variant fixture design system (Nee *et al.*, 1992).

Artificial intelligence technique, such as the expert system, has also been widely explored (Ferreira *et al.*, 1986). For instance, "AUTOFIX" (Pham and Lazaro, 1990) was an automated CAD system for configuring fixtures from a database of modular elements. The first complete and comprehensive expert fixture design system, which integrated set-up planning and fixture design for an automated manufacturing environment, was then developed (Senthil kumar *et al.*, 1992). A knowledge-based approach was also applied to develop an automated modular fixture design and assembly system (Dai *et al.*, 1996).

The methodology for automated configuration of fixturing points was developed in some of the reported works, such as the method of using finite element and heuristic rules of forces and kinematics law (Menassa and DeVries, 1989 & 1991), the projective geometry approach (Trappey and Liu, 1993), and geometric reasoning algorithm (Trappey *et al.*, 1993). As most of the point determination algorithms were applied on the prismatic workpiece involving planar fixturing faces, a methodology for applying V-blocks and clamps to non-prismatic workpiece was then developed (Hou and Trappey, 1997) such that the cylindrical faces can be used for fixturing.

Comprehensive reviews on the contemporary development of computer-aided fixture design and its future trends and needs were presented by (Hargrove and Kusiak, 1994) and (Cecil, 2001). One of the future trends will be the development of a comprehensive and complete automated fixture design system which integrates the CAD/CAM/CAE tools, such as with NC programming schemes to establish a fixture design system which is capable of detecting interference between a cutting tool and fixture assembly.

Most of the works reviewed above considered only the one aspect of the whole fixture design process, either automatic point determination or generation of fixture element assembly. Besides that, very few systems have incorporated the functionality of detecting machining interference (Senthil kumar *et al.*, 2000). This leads to a gap between the fixture design and manufacturing operations where the aspect of cutter paths is not considered during the design stage. As a result, re-design cannot be avoided when the cutter is found to interfere with the fixtures in the manufacturing set-up.

In this chapter a CAD-based methodology is used to develop the integrated computer-aided modular fixture design system which is built on Unigraphics (UG) solid modelling environment. UG/Open API (previously called UG User Function), a software development tool provided by UG, is used to develop the system software. This enables the provision of a user-friendly and graphical user-interface (GUI) based design environment. The structure of the integrated computer-aided modular fixture design system is shown in Figure 2.1.

The system integrates three different types of fixture design methodologies, i.e. interactive, semi-automated and automated fixture design module, into a single environment (Kow *et al.*, 1999). The user has the option to choose the most suitable module based on the design needs. The interactive fixture design (IFD) module facilitates the designer in selecting the fixturing faces, points and elements before the system assembles them through the fixture assembly routines. To reduce the design lead-time, the semi-automated fixture design (SFD) module is employed to automate the selection of fixture elements through the approach of geometric-reasoning algorithms. With the development of the automated fixture design (AFD) module, fixturing points can be determined automatically based on fixturing heuristics rules. Candidate lists of points are generated after obtaining the workpiece geometric information and by applying the fixturing rules, the exact fixturing points are chosen from the lists. Hence, a complete automation is achieved from the automatic determination of fixturing points to the selection of fixture elements. To produce an interference-free design solution, the system is developed to check machining interference between the cutter path and fixture assembly, through the use of cutter swept volume approach. Integrating the interactive, semi-automated and automated fixture design methodology into a single environment, provides the flexibility in reaching a feasible design solution by switching between the various modules.

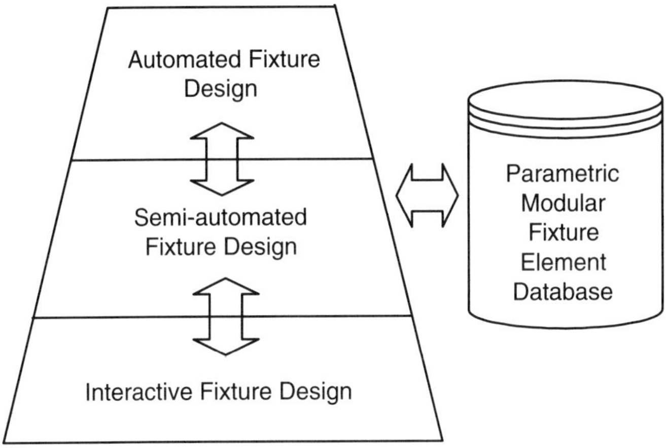

Figure 2.1 The structure of the integrated computer-aided modular fixture design system

2.2 Integrated Prototype System

2.2.1 Interactive Fixture Design System

The interactive fixture design (IFD) module was developed in UG, a commercial CAD system. By using UG/Open API, a programming interface tool provided by UG, fixture design algorithms are created and integrated to provide the necessary design functions and interfaces to aid the user in configuring a modular fixture set-up for a given workpiece. This facilitates the user in the selection of fixturing faces, points and elements through the developed user-interface menus. The design process is divided into the following sequences, *i.e.* loading of workpiece part file and fixture base, supporting, locating, clamping, and mounting elements. Mounting elements are used to raise the fixture elements above the fixture base plate. After loading the workpiece the user selects the appropriate base plate based on the size of the workpiece. The user then identifies the suitable fixturing points and elements which are loaded from the parametric modular fixture database constructed based on the standard fixture elements of IMAO Venlic BJS, a commercial hole-based modular fixturing kit (IMAO). The fixture elements are then positioned with respect to the workpiece in the correct orientation through the fixture assembling routines. During the design process, the fixture assembly tree structure is generated to categorize the fixture elements in the assembly based on the type of elements. Design modifications are provided by which fixture elements can be translated to a new position, rotated to a new orientation or removed from the assembly. The fully interactive nature of the module provides the necessary flexibility in achieving a custom design solution. The functions of the interactive fixture design module are summarized in Figure 2.2.

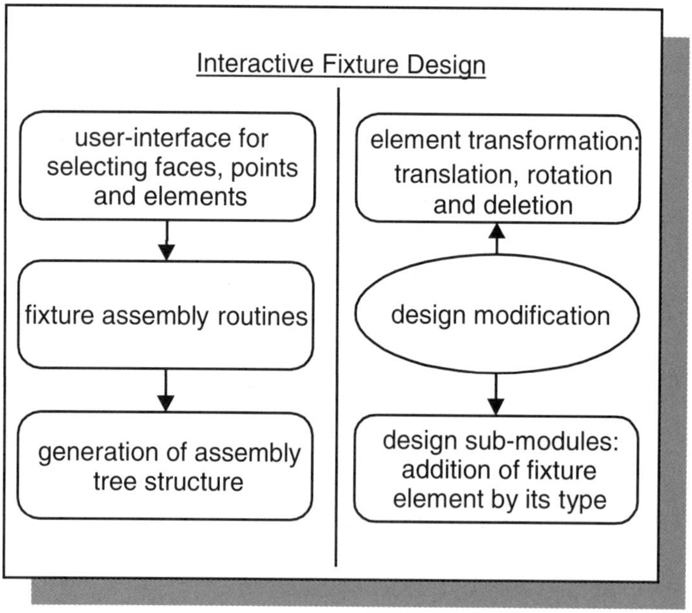

Figure 2.2 Interactive fixture design module

The Semi-automated Fixture Design (SFD) (Kow *et al.*, 1999) module aims at speeding up the fixture design process by automating the selection of fixture elements. Using the existing architecture of the IFD module, it automates the selection of fixture elements and maps the position of the elements onto one of the positioning holes of the fixture base. This eliminates the problem of the IFD module that the fixture elements may not be located exactly on the positioning holes. The user needs to input some information through the user-interface menu as in the IFD module, such as the fixturing faces, points, etc. to start the design process (Figure 2.3). Given the fixturing points, the exact location of those elements on the fixture base is determined through a method called "locating constraint mapping", which maps the fixturing points onto the positioning holes (grids) of a fixture base (Figure 2.4). In this method, a locating point is used as a reference point to map the subsequent fixturing points onto the nearest positioning grids based on the following equation characterized by a constant distance (50mm) between the positioning grids, *i.e.*,

$$\frac{x\_fixturing\_point - x\_reference\_point}{50.0} = n \tag{2.1}$$

$$\frac{y\_fixturing\_point - y\_reference\_point}{50.0} = n \tag{2.2}$$

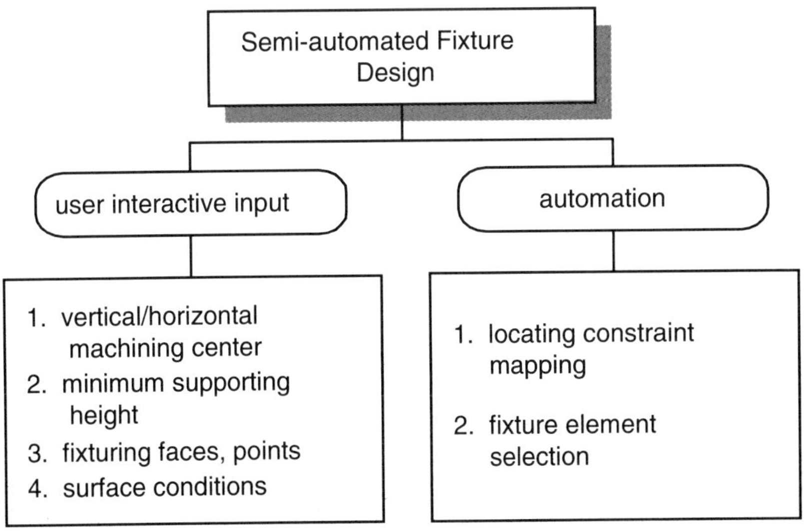

Figure 2.3. Semi-automated fixture design module

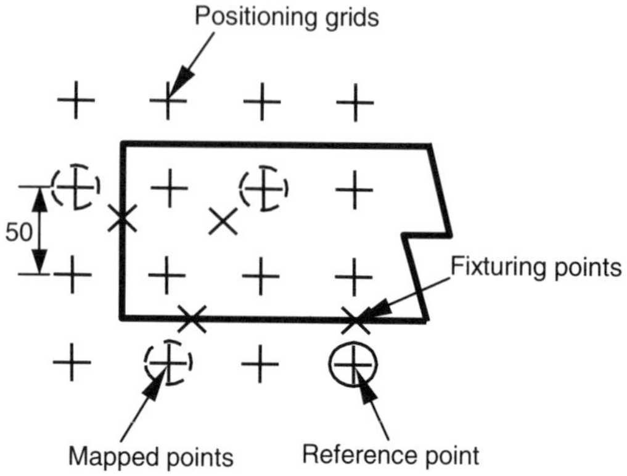

Figure 2.4. Mapping of fixturing points

Having determined the exact position of the fixture elements on the fixture base, the spatial information of fixturing points, such as their height from the base, and their horizontal distance from the fixturing face, are to be obtained automatically. This information is then processed by the rule-based geometric reasoning algorithms, which compare the fixturing spatial information with the geometric information of the fixture elements in the database and choose the most suitable fixture type and size, through sets of selection rules. An example of a typical selection rule is as follows:

```
if (( clamping height > 50 ) AND ( clamping height <= 70 ))
      then      {  select adjustable clamp type A;
                   find the exact size;  }
else if ( clamping height <= 50 )
      then      {  select adjustable clamp type B;
                   find the exact size;  }
else if ( clamping height > 70 )
      then      { select adjustable clamp type A;
                  find the exact size;
                  go to select adopter X;  }
```

A fixture tower can be formed by including the required number of mounting elements to raise the height of fixture element. Whether mounting elements are needed depends on the type of fixture element involved and the excessive height to be filled between the fixture element and the base.

2.2.2 Automated Fixture Design

The Automated Fixture Design (AFD) system is created to further enhance the efficiency of fixture design process and the quality of solution in terms of tool collision-free by adopting the existing interactive and semi-automated fixture design architecture. The framework of the developed AFD system is shown in Figure 2.5. The CAD environment and the various constraints are integrated using the application interface (API) provided within the UG CAD system. The fixturing elements are parametrically designed and stored as CAD models and when required, they are displayed and assembled automatically using the macros written in C language. Automated fixturing points determination module provided in the system performs an exhaustive search to reach a feasible design solution. Besides that, automatic detection is incorporated into the system to eliminate any interference caused by the cutter movement during machining or due to the close proximity of two fixture elements. Generally, there are two kinds of interference that can occur in fixture design, namely *design interference* and *machining interference*. The former is the interference among the fixture elements due to close proximity of two fixture elements. The latter is the interference between the fixture assembly and the planned tool path.

Machining interference detection plays an important role in efficient process planning. Having integrated the fixture design module with manufacturing planning, interference between the fixture set-up and the planned cutter path can be detected and rectified during the design stage to avoid tool collisions. Early detection of interference saves the time and cost of manufacturing.

2.3 Automated Fixturing Points Determination

To determine the fixturing points automatically, the user is required to identify the faces of the workpiece to be machined, as well as the locating and supporting faces. Suitable clamping faces will be determined automatically together with the corresponding clamping points based on the fixturing principles. Identified machining faces will be avoided from selecting them as clamping faces.

The 3-D model of the workpiece is projected onto a 2-D plane to form a projected envelope of the workpiece. This implies that only (x, y) coordinates of the geometric data are involved in the point determination. The z-coordinate of the points is obtained later by considering the workpiece geometry and the size of the chosen fixture element. This turns boundary faces of the workpiece into a series of connecting lines enclosing the interior faces as shown in Figure 2.6. Hence, the selected locating and supporting faces are transformed into boundaries and planes respectively. This reduces the complexity of the fixture configuration problem from a 3-D search operation to 2-D.

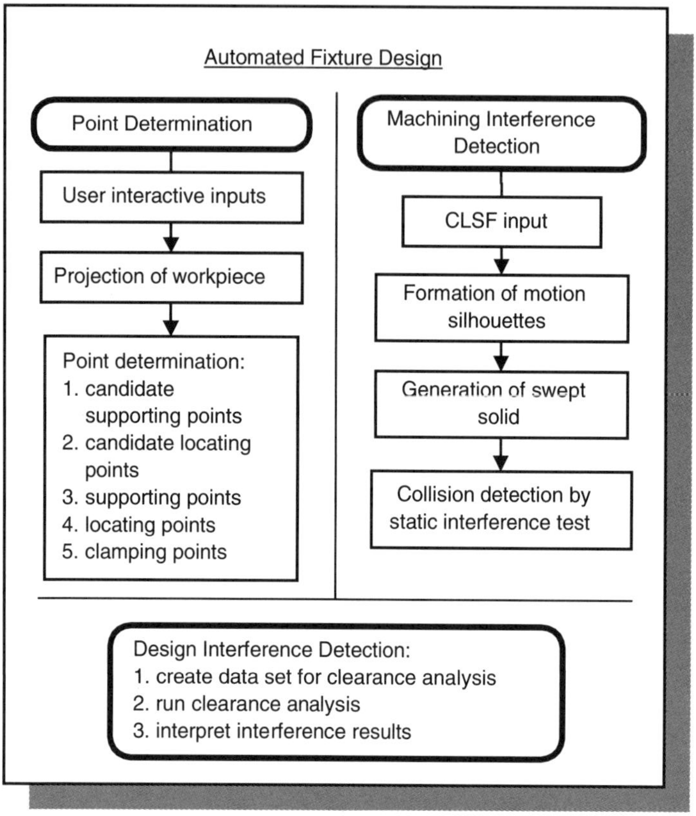

Figure 2.5 Automated fixture design module

Only boundary faces that are perpendicular and interior faces that are parallel to the fixture base can be used as the fixturing faces. Therefore, the selected locating and supporting faces must be perpendicular and parallel to the fixture base respectively. This is acceptable, as fixturing elements are generally not placed on inclined surfaces due to the tendency to slip and difficulty in maintaining the repeatability.

There are five stages in determining the fixturing points given the locating and supporting faces together with the machining faces: they are the generation of candidate supporting and locating points, selection of supporting, locating and clamping points. Candidate points are generated on the 2-D projected envelope of the workpiece using the fixed interval method as shown in Figure 2.6. Heuristics search on the list of candidate points in accordance to the fixturing rules is employed to determine the exact fixturing points.

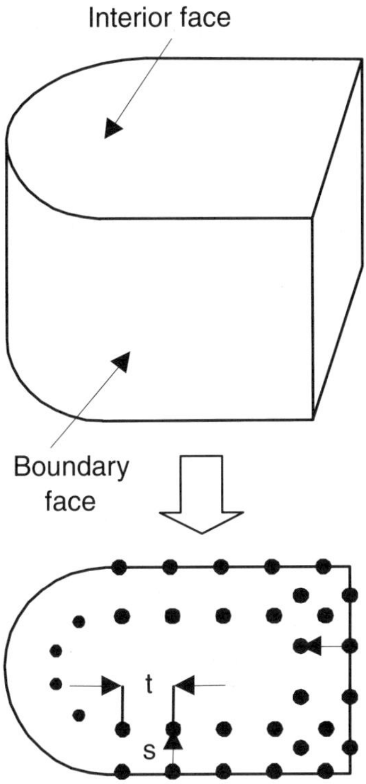

Figure 2.6 Projection of workpiece profile and generation of candidate points

2.3.1 Generation of Candidate Supporting Points

To generate the candidate supporting points, a fixed interval represented in the form $\{s, t\}$ is applied on to the outer boundary of the 2-D projected profile of the workpiece. This generates a list of candidate points around the outer boundary of the face, with an offset distance s from the boundary and constant distance t apart. s indicates the offset distance along the inner normal of an edge while t defines the interval distance between two adjacent points. The value of s and t are chosen considering the minimum radius value of standard support and the distance between the positioning holes of fixture base.

The procedures for generating the candidate supporting points are as follows:

1. The boundary edge data, such as vertices, centre and radius (if it is circular), and inner normal (if it is linear), are obtained;

2. Offset the vertices of each edge by a distance of *s*, along the inner normal of the edge; for a circular edge, its centre and radius will be used to calculate the inner normal at start and end point of the edge;
3. Find the intermediate points between each pair of the offset vertices at an interval *t*;
4. Store the intermediate points, together with the offset vertices as a list of candidate points;
5. Check if these points are contained within the boundary; discard points that are outside the boundary; and
6. Check if there is any duplication of points in the list; if yes, discard the duplicated points from the list.

An example of the generated candidate supporting points around a given boundary is shown in Figure 2.7.

2.3.2 Generation of Candidate Locating Points

The generation of the candidate locating points is similar to that of the generation of supporting points, except that no offset *s* is needed (refer to Figure 2.8). Locating faces are projected onto lines, represented by edge vertices between them. The intermediate points between each pair of vertices are determined at an interval *t* with offset distance *s* equal to zero. Only the intermediate points are stored as a list of candidate locating points because the potential locating points should not be located on the corner edge of the locating face.

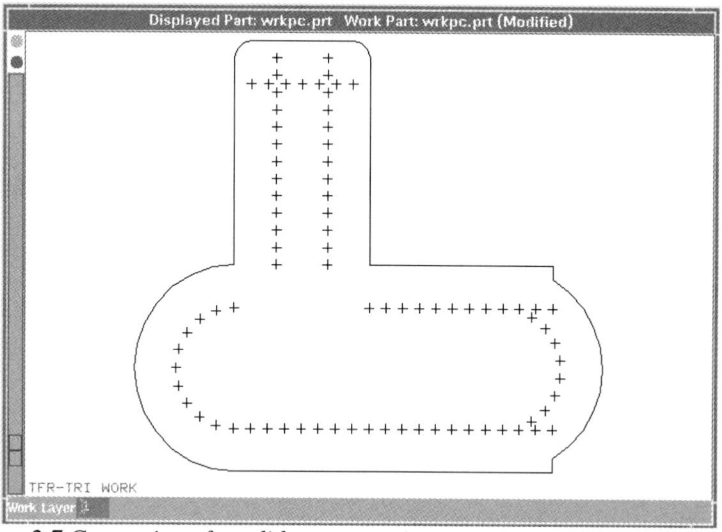

Figure 2.7 Generation of candidate supporting points around the boundary (top view)

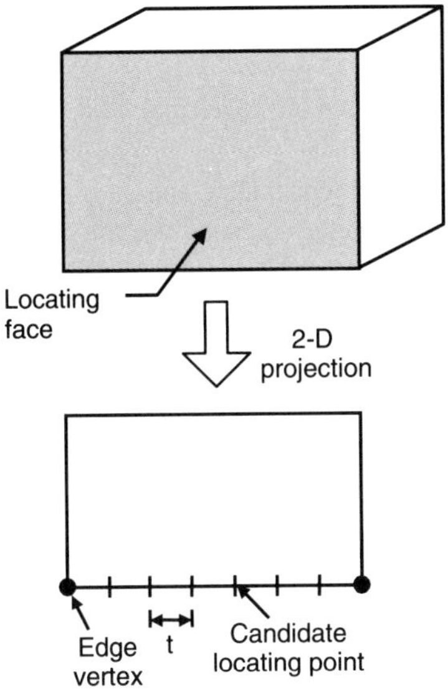

Figure 2.8 Generation of candidate locating points

2.3.3 Selection of Supporting Points

From the list of candidate supporting points, the exact supporting points are determined by evaluating the candidate points using equations and rules that are formulated from the fixturing principles. Stability constraint is an important factor in determining the supporting points. According to the 3-2-1 locating principle, at least three supporting points are required to maintain the stability of the workpiece. These supports must be placed as far apart as possible such that the mass centre of the workpiece is contained within the triangle formed by the three supports. All those triangles, which do not contain the mass centre of the workpiece, are eliminated and the vertices of the triangle (in this case, three points only) with the largest area are chosen as the supporting points.

The procedure for finding the supporting points with the widest placement is outlined in the following:

1. Find the total number of triangle combinations from the list of candidate points:

$$total\_combination = {}^{n+1}C_3 = \frac{(n+1)(n)(n-1)}{3 \times 2 \times 1} \quad (2.3)$$

where (n + 1) is the total number of points in the candidate list.

2. Find the triangle area for all the combinations using the formula as follows:

$$area(i,j,k) = \frac{1}{2}\left(X_iY_j - Y_iX_j + Y_iX_k - X_iY_k + X_jY_k - Y_jX_k\right) \quad (2.4)$$

where (i,j,k) is the vertices of the triangle and (X,Y) is their coordinate values;
3. Find the mass centre of the workpiece, and
4. Check if the selected triangle encloses the mass centre by using ray-crossing method. (Rogers and Adams, 1990)

Having found the (*x*, *y*) value of the supporting points, the *z*-coordinate of the supporting points is determined by projecting the point onto the surface of the supporting face.

2.3.4 Selection of Locating Points

The locating points are selected from the candidate locating points based on the principle of furthest distance such that the accuracy on the effect of workpiece irregularity is minimized to achieve an accurate location. Moreover, the locators are not positioned on opposite surfaces. This is to avoid the redundant location of the workpiece. In order to satisfy these requirements, the locating faces specified by the user are subject to the following conditions.

1. The number of locating faces must not exceed two so as to avoid a redundant location. Based on the 3-2-1 fixturing principle, there are two locating planes for accurate location, namely the secondary and tertiary plane, containing two and one locators respectively. Hence, a maximum number of two locating faces with the same surface normal must be selected to form the secondary locating plane while there is only one face for the tertiary locating plane. The locating faces are grouped into secondary and tertiary planes based on their projected length and direction. The face with a larger projected length is used as the secondary plane such that the locating pins can be placed as far apart as possible.
2. The secondary and tertiary locating planes must also satisfy the necessary angular constraint, which indicates that the angle between them must be within the range of 45° to 135°, and both faces must be perpendicular to the primary datum plane. This prevents the locators from being placed on opposite surfaces.

The two locating points on the secondary plane may be on the different faces but must have the same surface normal. The locating points are chosen from the candidate list such that they are located as far apart as possible, but not touching the corner edge of the face. The third locating point is selected such that it is nearest to the centre of the projected tertiary locating face.

Having found the (x, y) coordinate, the height (z-coordinate) of the locating points is initialised to the same level as the height of the centre of the locating face. A point containment routine is used to check if the points are within the boundary of the locating face. If a point is outside the boundary, then a list of boundary points with an offset distance from the boundary is generated and the point that is nearest to the off-sided point will be chosen as an alternative locating point.

2.3.5 Selection of Clamping Points

Usually, the clamping force is always directed towards the locators in order to force the workpiece to contact all locators. Therefore, there are a maximum of three top clamps against each support, and two side clamps against each locating plane respectively. However, generally it is not necessary to apply clamping against all the supports and locators. In this system, the required number and type of clamping depends on the availability of clamping faces, which in turn is affected by the number of faces to be machined. Whether a certain type of clamping can be applied depends on the availability of clamping points and the suitability of clamping elements. In view of this, selection rules are incorporated such that it prioritises the use of top clamping over side clamping, and suitable size of elements can be found for all the clamping points.

Top clamping

A candidate list of top clamping faces is obtained by considering the faces that are opposite to the supporting faces and those faces that are not machined. To obtain the suitable clamping points, first, the supporting points are projected upwards onto each candidate face by raising the z-coordinate of the supporting point to the height of each candidate face. Then, for each candidate face, the point containment routine is employed to check if the projected supporting point is on the face. It is assumed that the supporting point will hit either one candidate-face or not at all when projected. The candidate-face that contains the corresponding projected point will be chosen as the exact top clamping face. If there is more than one candidate-face that contains the corresponding projected point, only one candidate-face is arbitrarily chosen as the exact top clamping face. The corresponding projected supporting point will be the exact top clamping point.

Side clamping

Side clamping is used to hold the workpiece against the locators and a maximum of two side clamps are required to hold against each locating plane. Basically, side clamping faces and points are selected such that the clamping direction passes through the triangle formed by three locating points. Hence, the clamping face and point that are directly opposite to the locating plane are most preferred.

As in the case of top clamping, a list of candidate faces that are opposite to the locating planes and not to be machined is first generated. Then, the mid-point of the locating plane is projected along the surface normal of each candidate face onto the plane of the candidate face. For each candidate face, the point containment routine is employed to check if the corresponding projected point is on the face. The candidate face that contains the projected point will be chosen as the exact side clamping face while the projected point will be the exact side clamping point. If more than one candidate face is found suitable, then the face that is farthest from the locating plane is chosen.

When there is no directly opposite face that is suitable for clamping, a new list of candidate faces, comprising the available faces that are perpendicular to the supporting plane, will be generated. From the list of candidate faces, suitable clamping faces are determined considering the clamping direction passing through the triangle (locating triangle) formed by the three locating points with the smallest angle of diversion. The clamping direction is defined by the inverse surface normal of the clamping face whereas the angle of diversion is the smallest angle between the locating and clamping face normal. When the angle of diversion is zero, it implies that the locating and clamping faces are parallel to each other. From the refined list of candidate faces, the face with the smallest angle of diversion is selected as the exact clamping face, while the centre of the face is the clamping point.

2.4 Machining Interference Detection

The detection of machining interference is realized using the cutter swept volume approach based on the concepts of swept solids method (Ganter and Uicker, 1986). The swept solids method outlines three steps in generating the solid that traces the trajectory of a moving object, *viz.* motion silhouettes extraction, triangulation of silhouettes into a swept surface, and extraction of initial and final position information. The method has been modified in the present system to suit the situation of machining operations performed on a workpiece under the grip and location of a fixture assembly. The cutter swept volume approach is implemented considering the following four stages:

1. Cutter location source file;
2. Formation of motion silhouettes;
3. Generation of cutter swept solid;
4. Collision detection using static interference test.

2.4.1 Cutter Location Source File

In Unigraphics (UG), the cutter path information is stored in a text file, called Cutter Location Source File (CLSF), which can be obtained from UG/CAM before NC post-processing. The cutter path information is basically composed of operation name and type, operational parameters, tool data, the cutter motion and coordinates of its location. Subroutine is written to extract the operation type, tool geometry, cutter motion and location from the CLSF such that they can be used for generating the cutter swept solids.

2.4.2 Formation of Motion Silhouettes

Though the tool motion geometry can be in different shapes, in this chapter, it is simplified as cylindrical so that the geometry can be easily characterized by only two parameters, *i.e.* diameter and length. Motion silhouette is composed of a set of points whose motion is found to be tangent to the surface of a moving body. Since the tool geometry is generalized as cylindrical, the motion silhouette of a cutter can be represented by a motion sectional plane, which is formed by the points whose motion is tangent to the cutter periphery. The motion sectional plane is always perpendicular to the cutter trajectory at each cutter centre as shown in Figure 2.9.

Referring to Figure 2.9, the cutter moves along the horizontal direction where the cutter centres lie on the same plane. The cutter motion is restricted to linear or circular on the plane. To obtain the motion sectional plane, the points lying on the plane and the cutter periphery, *i.e.* A, B, C and D, must be found given the information of cutter location and its geometrical size. Then, the motion sectional plane is formed, by joining all the points with four lines. The motion sectional planes at all cutter centres will constitute the motion silhouettes of the cutter, tracing the cutter trajectory.

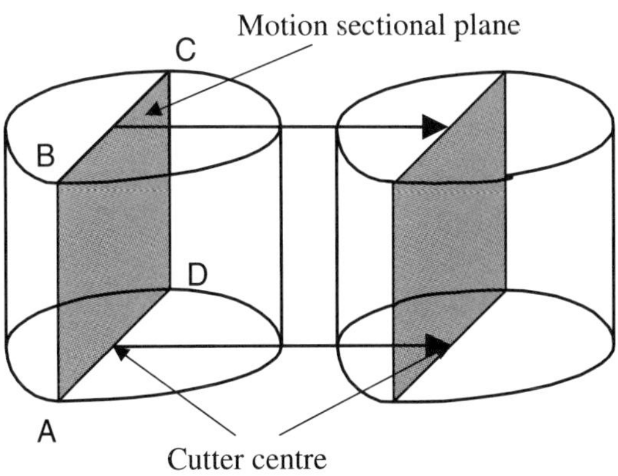

Figure 2.9 Generation of motion silhouettes

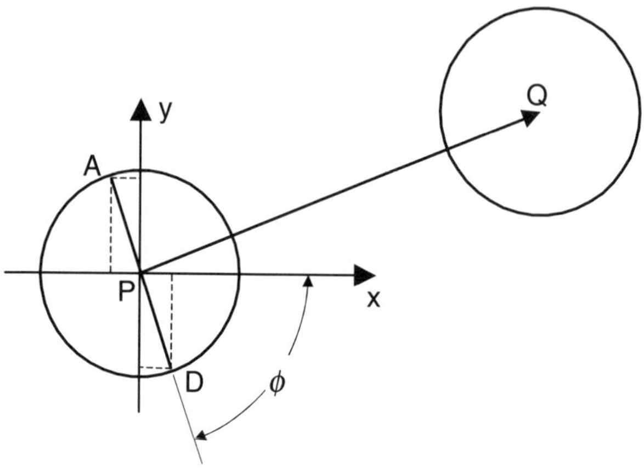

Figure 2.10 Linear motion of the cutter on the same plane

Points A and D on the bottom of the cutter are first obtained. Points B and C are then determined by raising A and D at the given length of the cutter. The procedures of finding the coordinates of A and D are described in the following steps:

Case (I): Linear motion (refer to Figure 2.10)

1. Find vector $\overrightarrow{PQ}$. Since the cutter centres P and Q are on the same plane, only x- and y-coordinates are considered.
2. Find a vector normal to $\overrightarrow{PQ}$. Such that $\overrightarrow{PQ} = \langle x \quad y \rangle \quad \therefore \overrightarrow{AD} = \langle -y \quad x \rangle$
3. Find the angle, ϕ, between $\overrightarrow{AD}$ and X-axis of the coordinate system within the range of 0° to 90°; and
4. Obtain points A and D based on the two cases of ϕ: (a) counter-clockwise or, (b) clockwise from X-axis. Their coordinate values are confirmed by checking if their vectors equal $\overrightarrow{AD}$.

   Case (a): $A = \left(P - \frac{d}{2} cos\phi \text{ , } P - \frac{d}{2} sin\phi \right) \quad D = \left(P + \frac{d}{2} cos\phi \text{ , } P + \frac{d}{2} sin\phi \right)$

   Case (b): $A = \left(P - \frac{d}{2} cos\phi \text{ , } P + \frac{d}{2} sin\phi \right) \quad D = \left(P + \frac{d}{2} cos\phi \text{ , } P - \frac{d}{2} sin\phi \right)$

   where d is the diameter of the cutter.

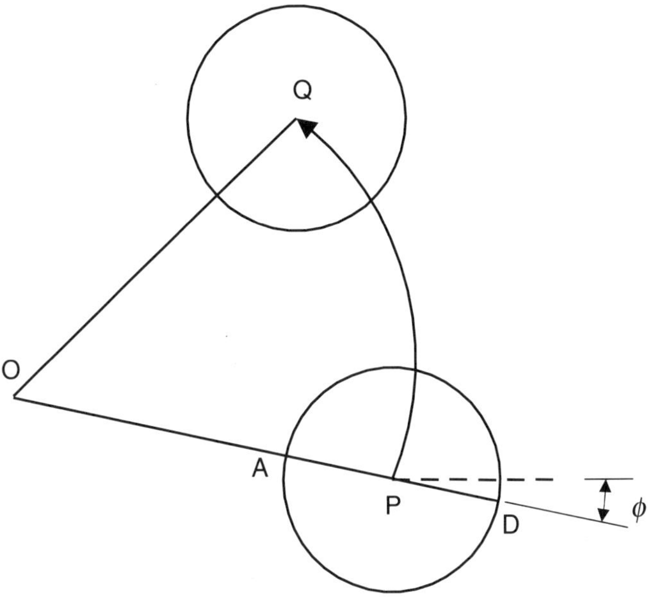

Figure 2.11 Circular motion of the cutter on the same plane

Case (II): Circular motion (refer to Figure 2.11)

1. Find the vector $\overrightarrow{OP}$ where O is the centre of circular motion.
2. Find the angle ϕ, and
3. Obtain the points A and D as in previous Case (I).

Apart from planar movement, the cutter can also traverse linearly between planes in different heights as shown in Figure 2.12. It can be along Z-axis (vertical movement) or at an angle to Z-axis (inclined movement). For inclined movement, the procedures of finding A, D, C and B, and then forming the motion sectional planes are the same as above except that the vector PQ is projected into a 2-D vector, ignoring the *z*-coordinate. For vertical movement, the motion sectional plane is defined directly by creating a circular curve of the same diameter as the cutter diameter and centred at the cutter centre.

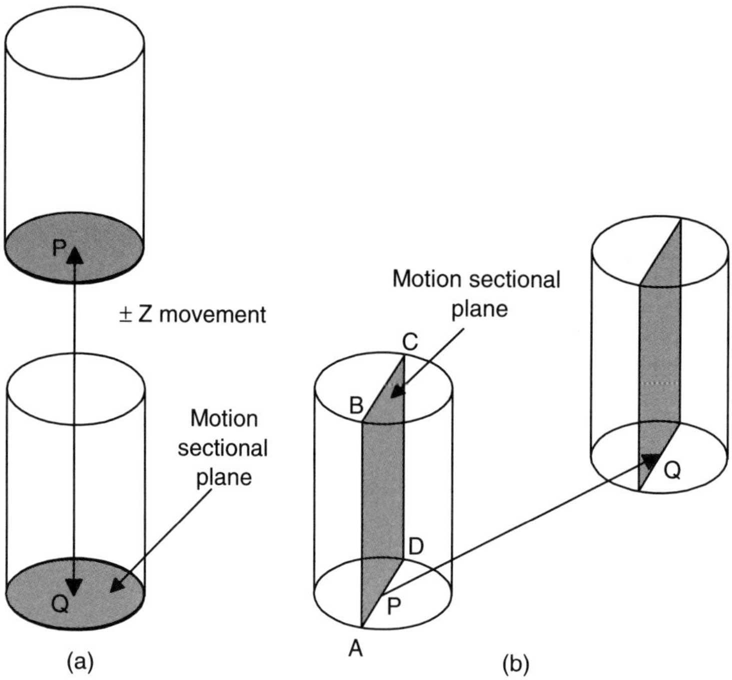

Figure 2.12 (a) vertical movement (b) inclined movement

2.4.3 Generation of Cutter Swept Solid

The task of generating swept solid across the set of motion silhouettes is analogous to the solid modelling problem of constructing a solid object from a planar profile. Hence, given the motion sectional planes at every cutter location, the cutter swept solid is created through extruding or revolving the sectional planes along the cutter trajectory. The swept solid of a linear motion is accomplished through the extruding operation while the circular motion is through the revolving operation. To complete the body of cutter swept solid, a cylinder is created at every cutter location. An example of cutter swept solid generated from a profiling motion of a milling tool is shown in Figure 2.13.

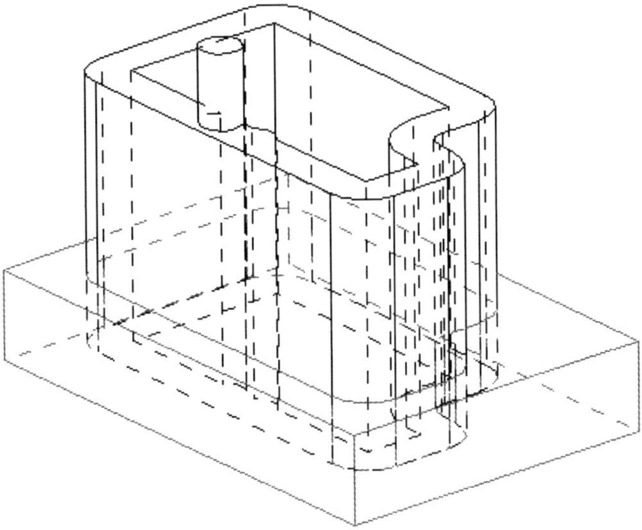

Figure 2.13 Cutter swept solid of a profiling operation

2.4.4 Collision Detection by Static Interference Check

The use of cutter swept solid has transformed the dynamic interference of cutter motion into static interference. Tool collision is implied by the existence of a volume of interference between the tool and the desired fixture. The static interference detection between cutter swept solid and the fixture assembly is performed in a manner similar to design interference detection. By storing the cutter swept solid as a component part of the assembly, the interference check is accomplished through the function of clearance analysis in the UG assembly module.

When interference occurs, the interfered fixture element will be highlighted and interactive rectification actions, such as translation, rotation, or replacement of a new fixture element, will be applied to correct the collision error. If the location of interfered elements is vital to the machining operation and cannot be changed, then a new CLSF that defines the new cutter path to avoid collision has to be created. Therefore, with the aid of the developed machining interference detection, fixture design produced by the developed system is free from tool collision error, so is the NC program generated from the CLSF.

2.5 Illustrative Example

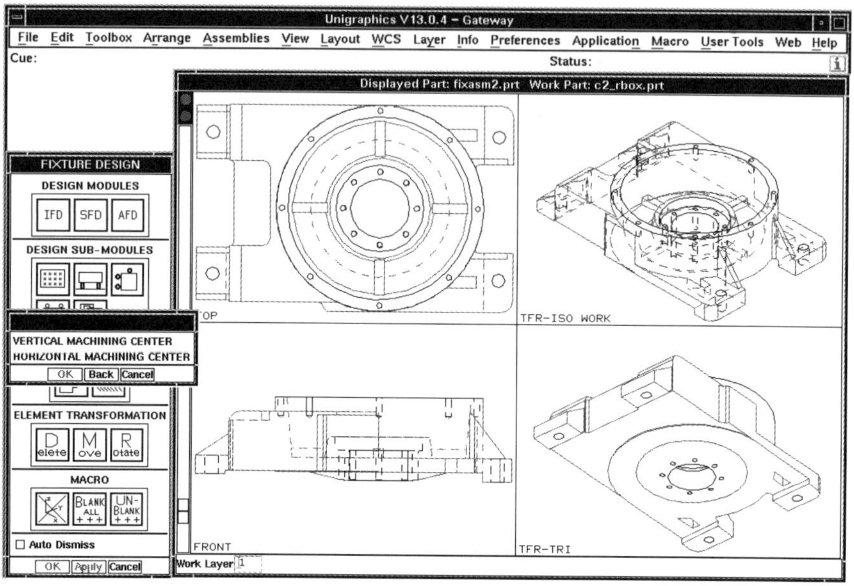

Figure 2.14 An example workpiece shown in various views

The developed interference-free fixture design and assembly system is explained with a case study. A fairly complex 3-D workpiece to be machined is shown in Figure 2.14. Face milling is to be performed on the cylindrical top face, then followed by finishing the counter-bore hole. Lastly, all the small holes are to be drilled. The automated fixture design (AFD) module has been applied to reach a feasible design solution efficiently without requiring the knowledge of fixturing points and elements from the user. The design process is initiated by the specification of initial inputs such as the workpiece part name, the type of machining centre, minimum supporting height and machining faces. Then, locating and supporting faces, with their surface conditions, are specified based on the user's knowledge on the workpiece reference datum and the faces suitable for fixturing. Given the locating and supporting faces, the locating, supporting, and clamping points will be determined accordingly based on the fixturing heuristic rules. The 3-2-1 locating principle is being followed where the supporting and locating points are located as furthest apart as possible in their respective plane. Top clamping is applied against all the supports according to the clamping priority selection rule. Then, the mapped positions of those points are obtained so that the fixture elements will be located on the positioning holes of the fixture base. Using geometric reasoning algorithm, appropriate fixture elements are selected to fulfil the fixturing spatial requirement at various points. To complete the fixture configuration, a base plate is added and grid points are created to represent the fixture positioning holes. The design solution produced by the AFD module is illustrated in Figure 2.15, which includes three supporting, three locating and three clamping elements.

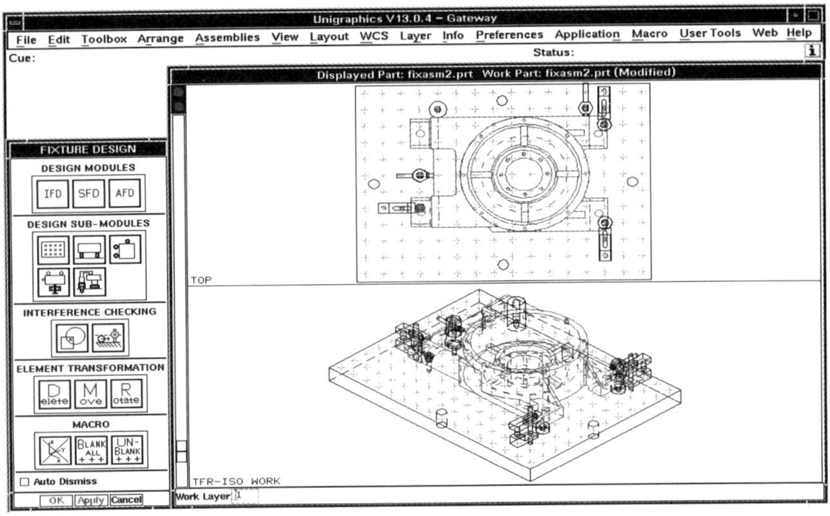

Figure 2.15 Fixture design and assembly from the AFD module

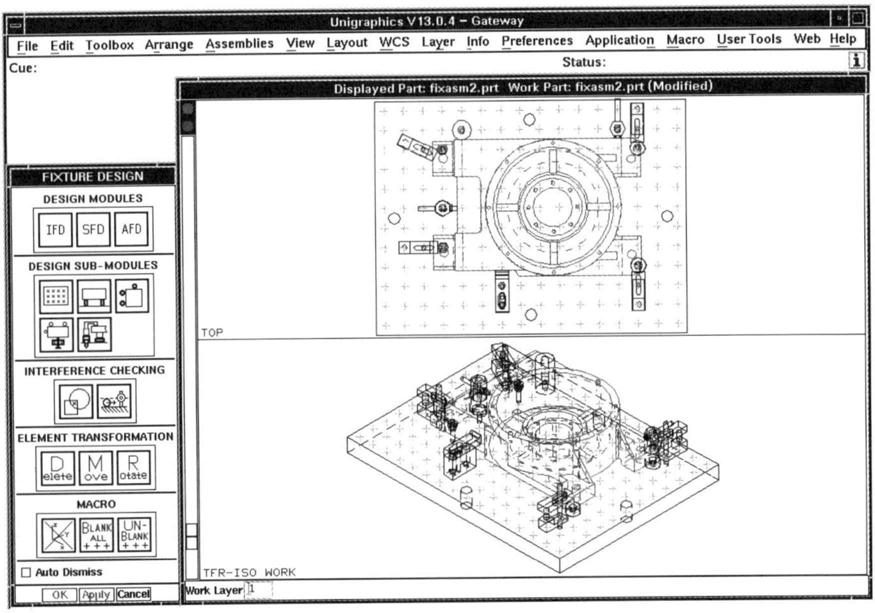

Figure 2.16 Addition of fixture elements using the IFD module

To better locate and hold the workpiece, additional fixture elements are required. Using the interactive fixture design (IFD) module, an adjustable support, top and side clamps are added. A rack block is added to mount the side clamp on the base plate (see Figure 2.16).

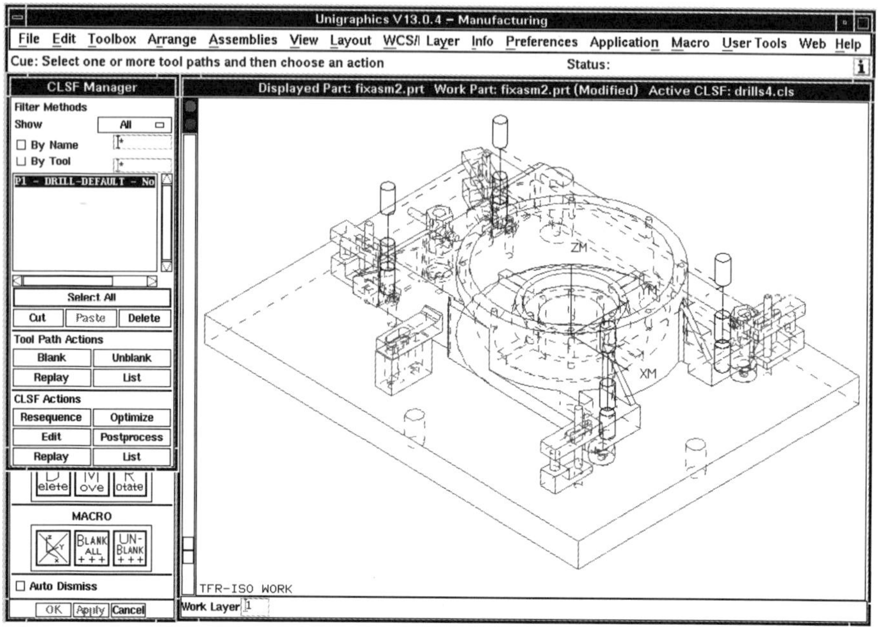

Figure 2.17 Simulation of cutter path in UG/CAM

Having obtained the fixture design and assembly, the machining operations to be performed are defined and simulated using UG/CAM. In Figure 2.17, the drilling operation to produce four holes of same diameter is defined and the cutter path is simulated and stored in a cutter location source file (CLSF). Given the cutter path, the next task is to determine if the fixture assembly and the cutter path interfere with each other. This is accomplished through the machining interference detection sub-module. The CLSF is read and processed by the sub-module to generate the cutter swept solid representing the space volumetrically swept by the motion of the drilling tool. Next, static interference check is performed to detect any interference. It is found that there is a tool interference between the tool path and the adjustable supports as shown in Figure 2.18. The interfering element is then highlighted for interactive rectification action. The situation is rectified by translating the support to the adjacent grid point that is still within the supporting face. Care is taken such that the stability of the workpiece is maintained. The final interference-free modular fixture assembly considering the NC tool path motions is shown in Figure 2.19 and Figure 2.20.

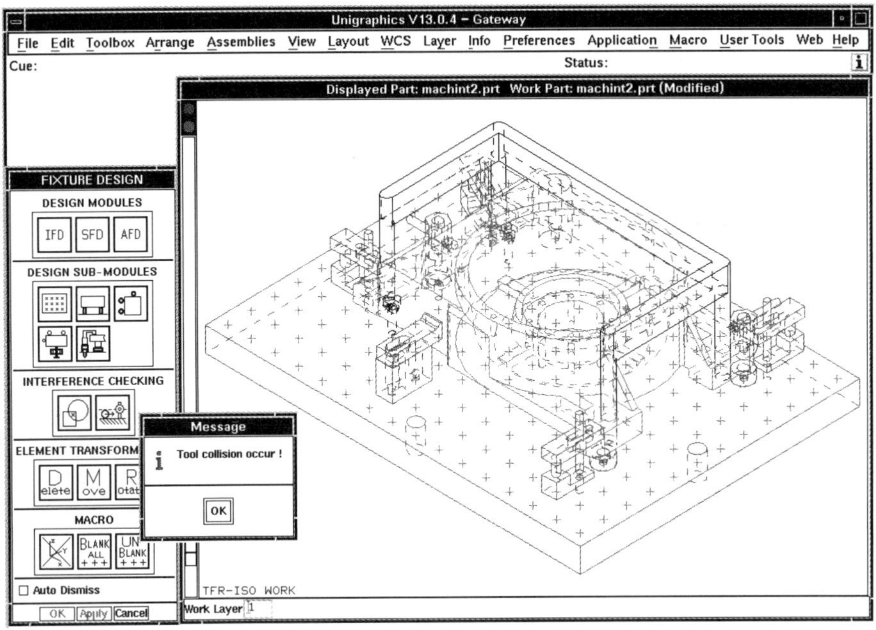

Figure 2.18 Interference test between cutter swept solid and fixture assembly

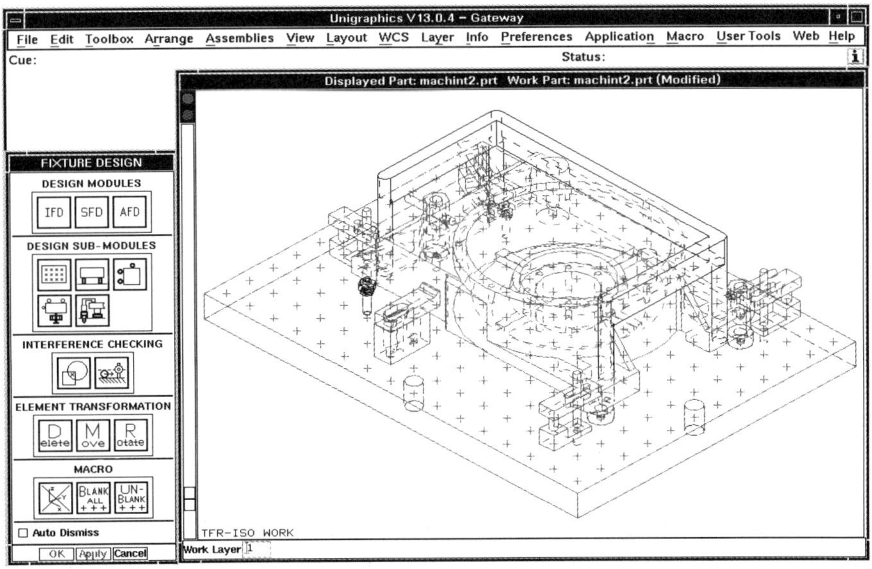

Figure 2.19 Modified fixture assembly that is free from tool collision

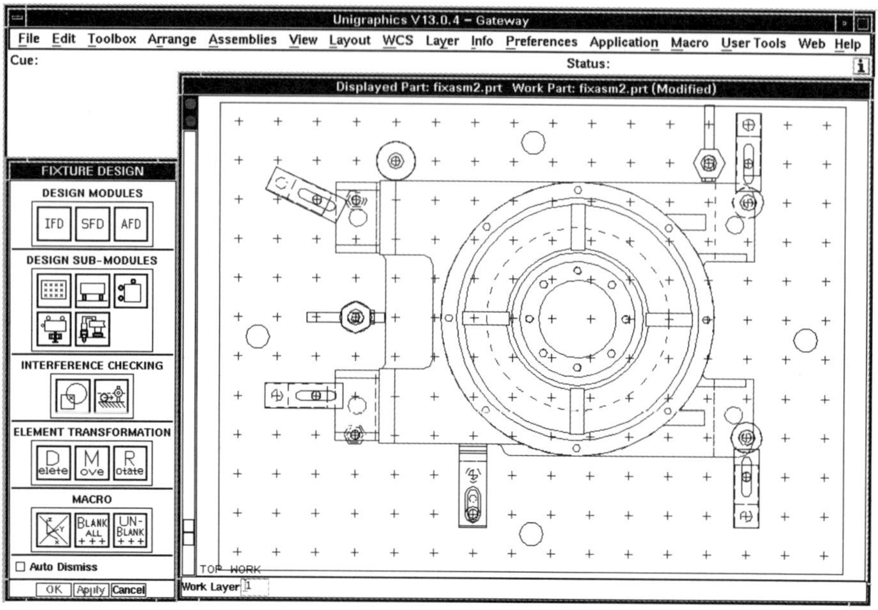

Figure 2.20 Top view of the final fixture assembly

2.6 Summary

This chapter reports the development of a novel fixture design system that is able to generate collision-free fixturing solution. The three systems, *i.e.* interactive, semi-automated and automated are integrated to form a complete modular fixture design and assembly system, interfacing with a parametric modular fixture database. The developed system is integrated with an NC tool path generation software such that an interference-free fixture design and assembly can be made efficiently for each workpiece set-up during process planning. It is flexible and robust enough for designing fixtures for complex workpieces. Work is currently being carried on to integrate the system with a tolerance analysis module such that the datum and fixturing faces can be identified automatically based on machining features.

2.7 Bibliography

Bai, X. and Y. Rong. Establishment of Modular Fixture Element Assembly Relationship for Automated Fixture Design. *Manufacturing Science and Engineering*, MED-Vol. 2-1, pp. 805-816. 1995.

Bausch, J.J. and K. Youcef-Toumi. Computer Planning Methods for Automated Fixture Layout Synthesis. *Proceedings of Manufacturing International'90*, Vol. 1, pp. 225-232. 1990.

Cecil, J. Computer-aided Fixture Design – a Review and Future Trends. *Int. J. Adv. Mfg. Technol.*, Vol.18, pp.790-793. 2001.

Dai, J.R., A.Y.C. Nee, J.Y.H. Fuh and A. Senthil kumar. An Approach to Automating Modular Fixture Design and Assembly. *Proc. Instn. Mechanical Engineers*, 211(Part B), pp. 509-521. 1996.

Darvishi, A.R. and K.F. Gill. Knowledge Representation Database for the Development of a Fixture Design Expert System. *Procs. of IMechE*, 202(B1), pp. 37-49. 1988.

Dong, X., W.R. DeVries and M.J. Wozny. Feature-based Reasoning in Fixture Design. *Annals of the CIRP*, 40(1), pp. 111-114. 1991.

Ferreira, P.M., B. Kochar, C.R. Liu and V. Chandru. AIFIX: An Expert System Approach to Fixture Design. *ASME Computer Aided/Intelligent Process Planning*, PED-19, pp. 73-82. 1986.

Gandhi, M.V. and B.S. Thompson. Automated Design of Modular Fixtures for Flexible Manufacturing Systems. *Journal of Manufacturing Systems*, 5(4), pp. 243-251. 1986.

Ganter, M.A. and J.J. Uicker. Dynamic Collision Detection using Swept Solids. *Journal of Mechanisms, Transmission and Automation in Design*, 108, pp. 549-555. 1986.

Gologlu, C. Machine Capability and Fixturing Constraints-imposed Automatic Machining Set-ups Generation. *Journal of Materials Processing Technology*, Vol.148, No.1, pp.83-92. 2004.

Grippo, P.M., M.V. Gandhi and B.S. Thompson. The CAD of Modular Fixturing Systems. *International Journal of Advanced Manufacturing Technology*, 2(2), pp. 75-88. 1987.

Hargrove, S.K. and A. Kusiak. Computer-aided Fixture Design: a Review. *International Journal of Production Research*, 32(4), pp. 733-753. 1994

Hou, J.L. and A.J.C. Trappey. A Methodology for Applying V-Blocks and Clamps to Non-Prismatic Workpart Fixtures. *International Journal of Computer Applications in Technology*, 10(3/4), pp. 152-169. 1997.

IMAO. *IMAO Venlic Block Jig System (BJS) - Solution to Quick, Accurate and Flexible Fixtures Assembly*. IMAO Corporation.

Kakish, J., P.L. Zhang and I. Zeid. Towards the Design and Development of a Knowledge-based Universal Modular Jigs and Fixtures System. *Journal of Intelligent Manufacturing*, Vol.11, No.8, pp.381-401. 2000.

Kang, Y., Y. Rong, J. Yang and W. Ma. Computer-aided Fixture Verification. *Assembly Automation Journal*, Vol. 22, No.4, pp.350-359. 2002

Kow, T.S., A.Senthil kumar and J.Y.H Fuh. An Integrated Approach for Collision Free Computer-aided Modular Fixture Design. *International Journal of Advanced Manufacturing Technology*, 16, pp.233-242,2000.

Ma, W., J. Li and Y. Rong. Development of Automated Fixture Planning Systems. *Int. J. of Advanced Manufacturing Technology*, 15:171-181. 1999.

Markus, A., Strategies for the Automated Generation of Modular Fixtures. *Symposium on Manufacturing Systems Design, Integration and Control, ASME Proceedings of Manufacturing International'88*, Vol. 3, pp. 97-103. 1988.

Menassa, R.J. and W.R. DeVries. Optimisation Methods Applied to Selecting Support Positions in Fixture Design. *ASME Journal of Engineering for Industry*, 113(11), pp. 412-418. 1991.

Menassa, R.J. and Warren R. DeVries. Locating Point Synthesis in Fixture Design. *Annals of the CIRP*, 38(1), pp. 165-169. 1989.

Nee, A.Y.C., A. Senthil kumar, S. Prombanpong and K.Y. Puah. A Feature-based Classification Scheme for Fixtures. *Annals of the CIRP*, 41(1), pp.189-192. 1992.

Nee, A.Y.C., K. Whybrew and A. Senthil kumar. *Advanced Fixture Design for FMS*. London: Springer-Verlag. 1995.

Pelinescu, D.M. and Michael Wang. Multi-objective Optimal Fixture Layout Design. *Robotics and Computer-Integrated Manufacturing*, Vol.18, Issues 5-6, pp.365-372. 2002.

Pham, D.T. and A. De Sam Lazaro. AutoFix - An Expert CAD System for Jigs and Fixtures. *International Journal of Machine Tools and Manufacturing*, 30(3), pp. 403-411. 1990.

Rogers, D.F. and J.A. Adams. *Mathematical Elements for Computer Graphics*. McGraw Hill Publishing Company. 1990.

Senthil kumar, A., A.Y.C. Nee and S. Prombanpong. Expert Fixture-Design System for an Automated Manufacturing Environment. *Computer-Aided Design*, 24(4), pp. 316-326. 1992.

Senthil kumar, A., J.Y.H Fuh and T.S. Kow. An Automated Design and Assembly of Interference-Free Modular Fixture Set-up. *Computer-Aided Design*, 32, pp.583-596. 2000.

Trappey, A.J.C. and C.R. Liu. Automated Fixture Configuration using Projective Geometry Approach. *International Journal of Advanced Manufacturing Technology*, 8, pp. 297-304. 1993.

Trappey, A.J.C. and S. Matrubhutam. Fixture Configuration using Projective Geometry. *Journal of Manufacturing Systems*, 12(6), pp.486-495. 1993.

Trappey, A.J.C., C.S. Su and S. Huang. Methodology for Location and Orientation of Modular Fixtures. *ASME Manufacturing Science and Engineering*, PED-Vol. 64, pp. 333-342. 1993.

Wu, Y., Y. Rong, W. Ma and S.R. LeClair. Automated Modular Fixture Planning: Accuracy, Clamping, and Accessibility Analyses. *Robotics and Computer-Integrated Manufacturing*, Vol.14, No.1, pp.17-26. 1998.

Chapter 3

Fixture Clamping Layout Synthesis

3.1 Introduction

Fixture design is a critical activity in the context of computer-integrated manufacturing. Manual design of a suitable arrangement of the fixturing elements requires the expertise and experience of the tool designer and it is a time-consuming procedure. Therefore, automatic fixture configuration is an important task to be addressed in manufacturing.

After a workpiece is located by a set of properly positioned locators, a critical problem arises as where to place the clamps so that force closure, and hence workpiece stability, can be guaranteed when external dynamic cutting forces are imposed on the workpiece. This chapter is dedicated to the problem of generating the best clamping points and positive clamping sequence for an arbitrarily-shaped workpiece for a machining operation.

Detailed review of the state-of-the-art on fixturing design automation (Hazen and Wright, 1990; Hargrove and Kusiak, 1994) revealed that most of the current automated fixturing systems are based on expert systems. The expert system approach relies upon existing heuristics and one of its limitations is that, without qualitative reasoning and geometric analysis, only generic types of fixturing elements, locating and clamping faces can be proposed. A geometrically constrained optimal fixturing design cannot be automatically generated. Therefore an analytical methodology for enumerating all the possible fixturing configurations and choosing a suitable configuration in terms of fixturing positions, clamping sequence and clamping intensities is yet to be developed.

In this chapter, as an effort towards achieving the aforementioned objective, an optimal clamp planning approach for arbitrarily-shaped workpieces based on computational geometry methodology is developed. A clamping analysis algorithm drawing on the metric of force closure is presented, in which all feasible clamping points are automatically detected by examining the convex hull of the bounding wrenches associated with constraining contacts. Optimal clamping points are then picked from the feasible clamping regions according to a proposed criterion based on the geometrical property of the convex hull. Clamping sequence is in turn decided by the areas of the feasible clamping regions where the clamps are situated.

A clamping equilibrium measure is also proposed so that robust clamping layout can be generated. At the end of this chapter, case studies are included to demonstrate the effectiveness and capabilities of the developed methodology.

3.2 Importance of Analytical Approach

When designing a new fixture scheme, if all the possibilities of fixturing alternatives are not exhaustively explored, the designer may settle upon a sub-optimal design or even an ill-conceived invalid design. The importance of a systematic and analytical approach to fixture design can be highlighted by the following example.

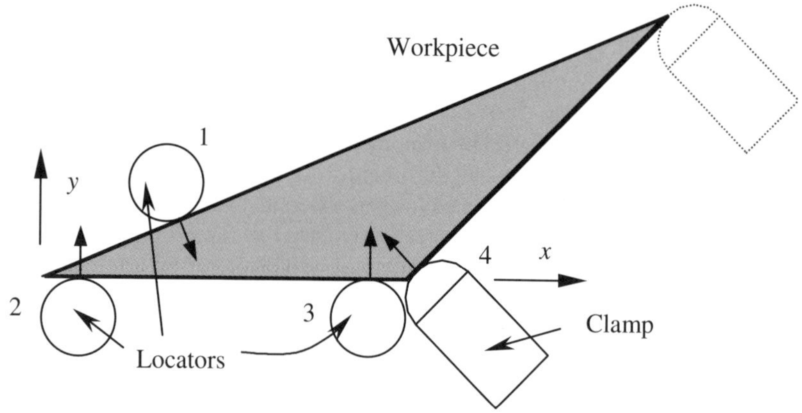

Figure 3.1 An example of an invalid fixturing scheme

An isosceles triangular workpiece as shown in Figure 3.1 is constrained by three round locating pins and a spherically-tipped clamping element. It is assumed that there is no friction between the workpiece and the fixturing elements. A close analytical examination of the fixturing configuration reveals that wherever the clamp is placed on the third side of the workpiece, force closure can never be achieved. This means that the established position of the workpiece would easily be disrupted by external load. The mathematical method of verifying force closure will be elaborated in Section 3.8 of this chapter.

This problem is intrinsically intractable through merely changing the clamping position or intensity. Such a locating arrangement has to be dropped and a new configuration must be conceptualised. In this particular case, a slight shift of locator No.1 from its current position to the middle of the edge would solve the problem.

3.3 Observation

The function of a locator is to provide the workpiece with deterministic location relative to the machine tool. Clamps are used to prevent any movement of the workpiece and ensure its contact with the locators during the machining process. From the view of kinematics, both locator and clamp can be regarded as restraining mechanisms that constrain the motion of the workpiece, and hence they are referred to as *restraints* in this chapter. The only difference is that the locators provide fixed contacts and hence are passive while the clamps can apply active force towards the inward normal direction of the contacting surface.

3.3.1 Clamping Equilibrium

The clamping operation is performed essentially to secure the workpiece before and during the machining process. When the clamps are activated, the forces exerted by the clamps should be counterbalanced by the resultant contacting force on the locators with no disturbance of the position and orientation of the workpiece. Such a clamping actuation may be referred to as achieving *clamping equilibrium*. It implies that all the forces at the contacting interfaces counterbalance each other. The self-weight of the workpiece is assumed to be small compared to the clamping force and is ignored here.

3.3.2 Force Closure

The applied clamping forces should be able to prevent any motion, whether infinitesimal or finite, of the workpiece under the influence of the variable cutting forces or any other external disturbances. Such a clamping conservation may be referred to as achieving *force closure*. Intuitively, a workpiece is said to have achieved force closure when any external load can be balanced by a non-negative linear combination of the contacting wrenches acting on it.

3.4 Contact Force Model

The immobilization of workpiece is achieved through a set of contacts between the surfaces of the workpiece body and the fixturing elements. The type of contact establishes the constraints on the forces that may be applied through the contact between two objects. It is assumed that the constraint is provided by hard point contacts and relies on friction. Both workpiece and fixturing elements are treated as rigid bodies. The force applied to the workpiece is restricted to be a pure force passing through the point of contact into the workpiece body, and it is bounded by the friction cone, with zero moment at the point of contact.

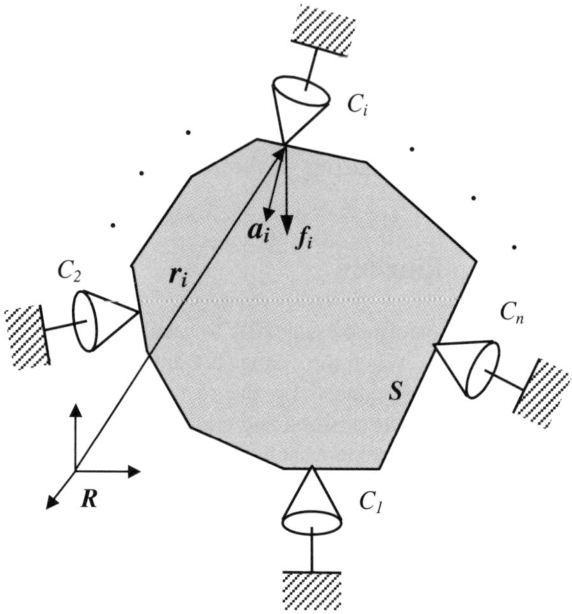

Figure 3.2 General workpiece and constraining contacts

Consider a general workpiece constrained by a set of fixturing elements as shown in Figure 3.2. Let the position of a contact point C_i be $\boldsymbol{r}_i = [\boldsymbol{x}_i \ \ \boldsymbol{y}_i \ \ \boldsymbol{z}_i]^T, \boldsymbol{i} = 1, 2, \cdots \boldsymbol{n},$ with respect to an arbitrary but fixed frame of reference R. Supposing the surface of the workpiece can be described by an implicit relation

$$\boldsymbol{S}(\boldsymbol{x}, \boldsymbol{y}, \boldsymbol{z}) = 0$$

where $(\boldsymbol{x}, \boldsymbol{y}, \boldsymbol{z})$ is a point in space defined with respect to R, and the surface has continuous first derivatives at the contact points, then the inner unit normal vector $\boldsymbol{a}_i$ at the contact point C_i is represented as

$$\boldsymbol{a}_i = \frac{\nabla \boldsymbol{S}(\boldsymbol{r}_i)}{\|\nabla \boldsymbol{S}(\boldsymbol{r}_i)\|} \tag{3.1}$$

where ∇ indicates the gradient operator and $\|*\|$ denotes the Euclidean norm of a vector.

Let $\boldsymbol{f}_i$ be the contact force exerted at the point of contact C_i by a restraint and μ_i be the coefficient of static friction, then $\boldsymbol{f}_i$ must satisfy the following constraint of the maximum static friction condition according to Coulomb's Law as shown in Figure 3.3(a):

$$\|\boldsymbol{f}_i\| \leq \sqrt{1+\mu_i^2}\,(\boldsymbol{f}_i \cdot \boldsymbol{a}_i) \qquad \text{for } i = 1, 2, ..., n \tag{3.2}$$

This is equivalent to that the contact force $\boldsymbol{f}_i$ must pass through C_i and lie in a friction cone centred about the internal surface normal at C_i where the tangent of its half-angle is the friction coefficient. In other words, the friction cone is a contact convex which describes the range of force that can be applied at the point. Obviously, the friction cone is an infinite convex cone. However, in the planar case, the friction cone degenerates into a friction sector that can be finitely generated by two vectors corresponding to its edges.

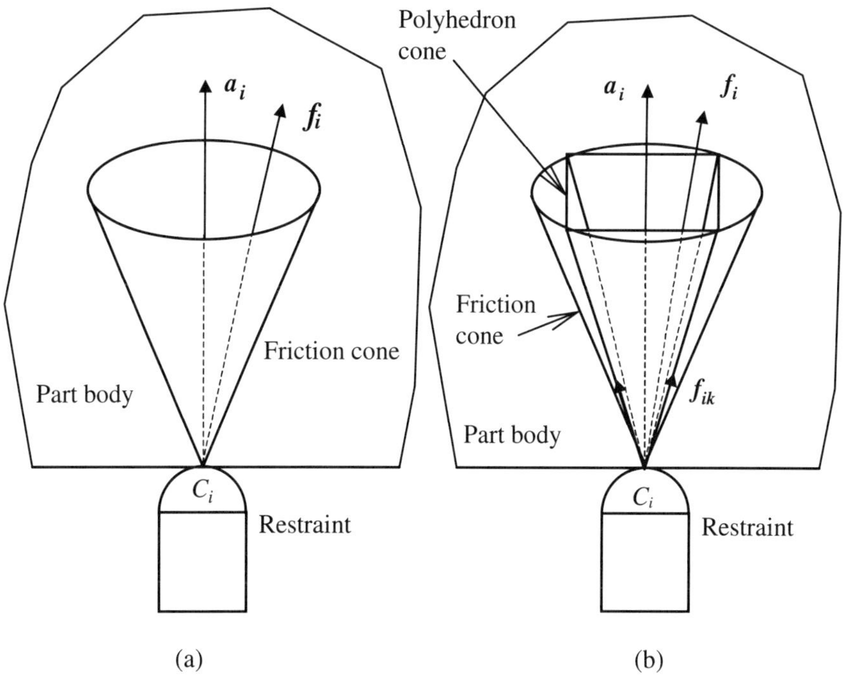

Figure 3.3 (a) Static frictional constraints for point contact and (b) polyhedral cone approximation

Since the contact force $\boldsymbol{f}_i$ exists only within the friction cone, it can be represented as a non-negative linear combination of an infinite number of generating vectors bounding the cone. For simplicity, the friction cone is usually approximated by a p-sided polyhedral convex cone (Kerr and Roth, 1986; Hirai and Asada, 1993) with typical values of p equal to 3, 4 or 6. The polyhedral cone is a finitely generated convex cone with a finite number of generators and is referred to as a pyramid as shown in Figure 3.3(b).

Let $\boldsymbol{f}_{ik}$ be the kth unit generating vector of a p-sided pyramid, which coincides with the edge of the polyhedron at the contact point C_i, then the contact force $\boldsymbol{f}_i$ can be represented as non-negative linear combination as follows:

$$\boldsymbol{f}_i = \sum_{k=1}^{p} \lambda_{ik} \boldsymbol{f}_{ik} \qquad \lambda_{ik} \geq 0 \quad \text{for k} = 1, 2, ..., \text{p} \tag{3.3}$$

This representation of the contact force $\boldsymbol{f}_i$ is in the wrench set associated with a finite number of wrenches of the generating vectors.

3.4.1 Conservative Representation and Accuracy

The truncated approximation to the friction cone is conservative in nature, and hence the inequality constraint expressed in Equation (3.2) is redundant. It can be proved that any contact force represented by Equation (3.3) must lie within the polyhedral cone. Since the pyramid is a convex cone, it is closed under multiplication by non-negative scalars. Geometrically, it means that if $\boldsymbol{P}$ is any vector in the cone, then every vector $t\boldsymbol{P}$, with $t \geq 0$, is also in the cone. Consequently for any convex combination

$$\frac{1}{\sum \alpha_k} \left(\alpha_1 \boldsymbol{f}_1 + ... + \alpha_p \boldsymbol{f}_p \right)$$

is in the cone, and so is the scalar multiple of

$$\sum \alpha_k \cdot \frac{1}{\sum \alpha_k} \left(\alpha_1 \boldsymbol{f}_1 + ... + \alpha_p \boldsymbol{f}_p \right) = \sum \alpha_k \boldsymbol{f}_k \, .$$

Therefore, the conservative nature is readily established by the fact that the truncated cone is a proper subset of the friction cone.

There is always a trade-off between the complexity and accuracy of the approximation. The approximation accuracy could be quantified by a coverage rate φ of the p-sided polygonal area over the total area as shown in Figure 3.4.

$$\varphi = \frac{p}{2\pi} \sin \frac{2\pi}{p} \tag{3.4}$$

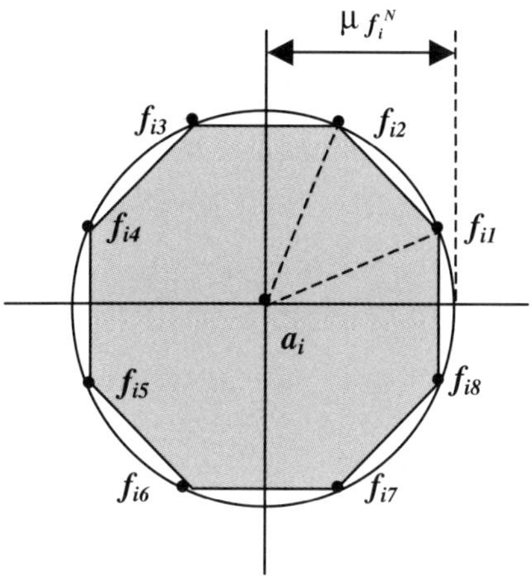

Figure 3.4 Orthogonal projection of friction cone onto the contact plane

3.4.2 Contacting Wrench

The normal forces f_i^N as well as friction forces f_i^T at the clamping and locating points can be derived from Equation (3.3) as follows:

$$f_i^N = \frac{1}{\sqrt{1+\mu_i^2}}\left(\sum_{k=1}^{p}\lambda_{ik}\right)a_i \tag{3.5}$$

$$f_i^T = f_i - f_i^N = \sum_{k=1}^{p}\lambda_{ik}\left(f_{ik} - \frac{a_i}{\sqrt{1+\mu_i^2}}\right) \tag{3.6}$$

The contact forces f_i are then resolved to an equivalent wrench w_i consisting of a force and moment about the origin of the reference frame R and defined as:

$$w_i = \begin{bmatrix} f_i \\ r_i \times f_i \end{bmatrix} = \begin{bmatrix} \sum_{k=1}^{p}\lambda_{ik} f_{ik} \\ \sum_{k=1}^{p}\lambda_{ik}(r_i \times f_{ik}) \end{bmatrix} \quad \text{where } \lambda_{ik} \geq 0 \quad \text{for k} = 1, 2, ..., \text{p} \tag{3.7}$$

The space of all the possible contact forces is a convex hull in d-dimensional space where d is the number of degrees of freedom of the workpiece. In the spatial fixturing configuration, the contacting wrench is a six-dimensional vector with three-dimensional translational force components and three-dimensional moment components. In the planar fixturing case, the wrench becomes a three-dimensional vector with two force components with respect to x- and y-axis and one moment component about z-axis as follows:

$$\boldsymbol{w}_i = \begin{bmatrix} \lambda_{i1}\boldsymbol{f}_{i1} + \lambda_{i2}\boldsymbol{f}_{i2} \\ \lambda_{i1}(\boldsymbol{r}_i \times \boldsymbol{f}_{i1}) + \lambda_{i2}(\boldsymbol{r}_i \times \boldsymbol{f}_{i2}) \end{bmatrix} \tag{3.8}$$

3.5 Gravitational Force

To be conservative, it is assumed that external disturbances could occur at any point during machining, and in any combination of directions and magnitudes. They may cause the processing forces to deviate from normal significantly. The deviation could potentially be quite large if the cutting tool is engaged with, say, the hardened material portion or foreign inclusions. Therefore fixturing stability is checked in the presumed absence of gravity in recognition that any instantaneous surge of processing forces might overwhelm the gravitational resistance.

For example, a milling operation causes a processing force to be imparted on the workpiece to the right as shown in Figure 3.5. The fixturing configuration is stable, only when the magnitude of milling force is confined to be less than the static friction induced by the gravity. This state of stability is detrimentally weak and limiting as it depends on the self-weight of the workpiece for obtaining force closure. To allow for the unpredictability of processing force fluctuation, the gravitational force is ignored in the construction of force-closure fixturing configuration. It is therefore obvious that one or more additional fixturing elements are needed for this particular set-up.

3.6 Force Closure Condition

Conditions for force closure have been studied extensively in the domain of robotic grasping (Lakshminarayana, 1978; Nguyen, 1988; Mishra *et al.*, 1987; Mishra and Silver, 1989; Ponce *et al.*, 1993) and are briefly introduced first for clarity. A computational geometry approach to the force closure condition with friction is then derived. Supplementary conditions are proposed for fixture verification in the next section.

In a positive grip model assuming non-frictional contacts, the wrench associated with each contact is defined as a primitive wrench as follows:

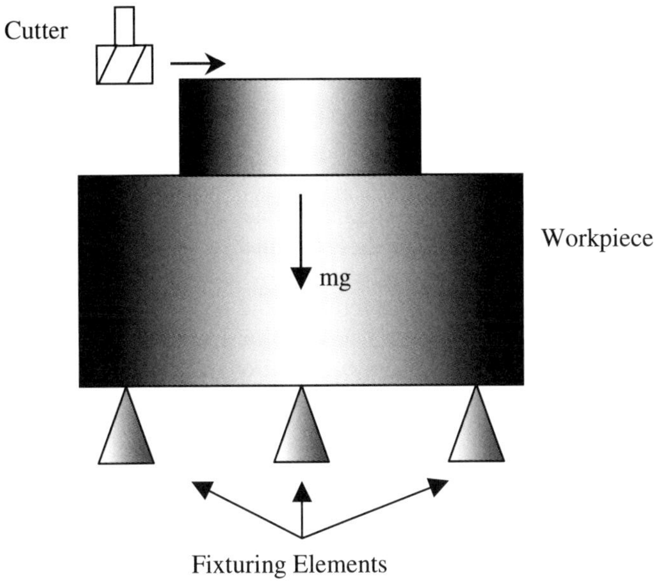

Figure 3.5 A fixturing setup for a milling operation which is conditionally stable

$$\boldsymbol{w}_i = \begin{bmatrix} \boldsymbol{a}_i \\ \boldsymbol{r}_i \times \boldsymbol{a}_i \end{bmatrix} \tag{3.9}$$

where $\boldsymbol{r}_i$ is the position of contact point and $\boldsymbol{a}_i$ is the inner unit normal vector.

A workpiece subjected to a set of primitive wrenches $\boldsymbol{w}_1, \boldsymbol{w}_2, \ldots, \boldsymbol{w}_n$ corresponding to a set of constraining contact points is said to have achieved force closure if and only if any arbitrary external wrench can be counterbalanced by varying the magnitudes of the wrenches subject to the constraint that the magnitudes are non-negative.

Geometrically, a system of n primitive wrenches $\boldsymbol{w}_1, \boldsymbol{w}_2, \ldots, \boldsymbol{w}_n$ is said to have achieved force closure if and only if the origin is in the interior of the convex hull of the wrench system in the Euclidean space $\Re^d$.

The presence of force closure is always desirable for a fixturing configuration to ensure workpiece stability when external cutting forces are imposed on the workpiece. The force closure definition implies that for any external wrench $\boldsymbol{w}_e = [\boldsymbol{f}_e, \boldsymbol{m}_e]^T$, the contacting wrench system is always capable of generating a negative counterbalancing wrench '-w_e' by exerting appropriate forces at the contact points.

In the presence of friction, a sufficient condition for achieving force closure is that there exists a set of forces in the *open* friction cones at the contacts such that the sum of the associated wrenches is zero, and is given as follows:

$$\sum_{i=1}^{n} \boldsymbol{w}_i = 0 \tag{3.10}$$

From the perspective of computational geometry, a sufficient condition of force closure can be derived next based on these established theorems.

As the friction cone can be approximated by a polyhedral cone, it is sufficient to detect the presence of the contact forces in the open truncated cones instead of the friction cones. Let $\boldsymbol{f}_i$ be a contact force in an open polyhedral cone, it can be represented by a strictly positive linear combination of its associated unit generating vectors $\boldsymbol{f}_{ik}$. Substituting Equation (3.7) into Equation (3.10) yields:

$$\sum_{i=1}^{n} \begin{bmatrix} \sum_{k=1}^{p} \lambda_{ik} \boldsymbol{f}_{ik} \\ \sum_{k=1}^{p} \lambda_{ik} (\boldsymbol{r}_i \times \boldsymbol{f}_{ik}) \end{bmatrix} = 0 \quad \text{where } \lambda_{ik} > 0$$

or

$$\sum_{i,k=1}^{n,p} \lambda_{ik} \overline{\boldsymbol{w}}_{ik} = 0 \qquad \text{where } \lambda_{ik} > 0 \tag{3.11}$$

where $\overline{\boldsymbol{w}}_{ik} = \begin{bmatrix} \boldsymbol{f}_{ik} \\ \boldsymbol{r}_i \times \boldsymbol{f}_{ik} \end{bmatrix}$ may be termed as the bounding wrench associated with the generating vector $\boldsymbol{f}_{ik}$ bounding the friction cone. Consequently, the contacting system achieves force closure if Equation (3.11) admits a strictly positive solution and the following proposition holds.

Proposition 3.1 In the presence of friction, a workpiece subjected to a set of constraining contacts is said to have achieved force closure when the origin lies strictly inside the convex hull of the associated bounding wrench vectors in $\mathfrak{R}^{\mathrm{d}}$.

The convex hull of the bounding wrenches $\overline{\boldsymbol{w}}_{11}, \overline{\boldsymbol{w}}_{12}, \ldots, \overline{\boldsymbol{w}}_{np}$ in $\mathfrak{R}^{\mathrm{d}}$ is the smallest convex polytope that contains all the wrenches and can be constructed using the Quickhull algorithm (O'Rourke, 1994; Barber *et al.*, 1996). Quickhull is a general dimension algorithm for computing the convex hull for a set of points. Direct application of this algorithm to a set of bounding wrench vectors in six-dimension is tractable.

3.7 Clamping Equilibrium

3.7.1 Grasping vs. Fixturing

Many of essential issues in fixturing are closely related to the theory of dextrous grasping in the robotics domain. While the objectives of both grasping and fixturing are identical, i.e. to immobilise an object so that certain manipulation could be carried out on it, they are fundamentally different in many aspects. The main difference between fixturing and grasping could be identified as follows:

- Only one, two or three fixturing elements are capable of applying active locking forces in workholding, while for robotic grasping, it is not unusual for all the fingers to exert restraining forces to the object, therefore more intricate force control algorithm is required;

- The accuracy requirement and ambient forces involved in workholding are rather stringent;

- In the fixturing context, it is almost always to employ hard contacts while the robotic fingers are often covered with rubber or other soft materials;

- The fixturing elements are geometrically more dedicated than the fingers used for robotic grasping.

3.7.2 Condition for Clamping Equilibrium

The concept of force closure implies that the mobility of the grasped part is completely restrained against external load by appropriately applying contacting forces at the restraints. The implicit assumption here is that the restraints are independently movable mechanisms that are all capable of exerting driving forces at the contacts. However, it is not the case when it comes to workholding operation where only clamps are able to apply active constraining forces while locators are simply passive and can only provide reaction forces. Therefore, supplementary conditions have to be imposed when the force closure is applied to fixture design as feasibility metric.

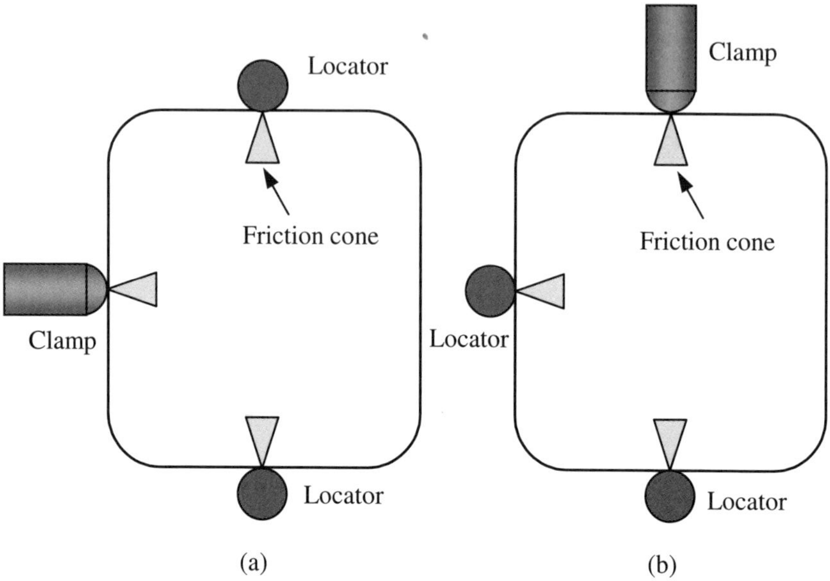

Figure 3.6 An example of a viable fixture configuration depending on clamping layout

Consider a 2-D fixturing example where a rectangular workpiece is secured by three frictional fixturing elements as shown in Figure 3.6. Friction cones are depicted as shaded sectors in the diagram. From the point of view of force closure, both fixturing configurations are identical and regarded as valid. However, it is intuitively obvious that, the fixture in Figure 3.6(b) can counterbalance arbitrary external forces acted on the workpiece by appropriately actuating the clamp, and the fixture in Figure 3.6(a) cannot resist any external force pulling the workpiece to the right horizontally, since the clamping force itself cannot be counterbalanced by the locators in the first place. Therefore, on top of force closure, a clamping equilibrium condition is needed for the existence of a valid fixturing configuration.

Proposition 3.2 In addition to the force closure requirement as defined in Proposition 3.1, a fixturing configuration is considered viable only when the *first* applied clamping force is counterbalanced by the contacting wrenches of the locators when all the restraints are replaced with frictionless contacts.

Consider a unit clamping force $\boldsymbol{a}_c$ and its corresponding wrench $\boldsymbol{w}_c$, as well as the contacting wrenches $\boldsymbol{w}_i$ with $\boldsymbol{i}=1,2,\cdots \boldsymbol{m}$ induced by the frictionless locating contacts, the condition is mathematically equivalent to the following equation admitting non-negative solution $\lambda_i \geq 0$:

$$\boldsymbol{w}_c + \sum_{i=1}^{m} \lambda_i \boldsymbol{w}_i = 0$$

or

$$\boldsymbol{w}_c + \boldsymbol{W}\lambda = 0 \tag{3.12}$$

where $\boldsymbol{w}_i$ is the primitive wrench as defined in Equation (3.9). The primitive wrench is the wrench of the unit normal force while the bounding wrench is the wrench of the unit generating vector of the polyhedral cone.

The clamping equilibrium requirement applies only to the first clamp exerted on the workpiece. The subsequent clamps, if any, need not be verified for clamping equilibrium because once the pre-loading between the interfaces of the workpiece and the restraints has been established by the first clamp, the other clamps would not disturb the workpiece in equilibrium.

The Linear Programming technique (Chvatal, 1983) can be employed to check whether the system of linear equations in Equation (3.12) admits the non-negative solution $\lambda \geq 0$.

Introducing two non-negative slack variables $\boldsymbol{p}$ and $\boldsymbol{q}$, a "Phase I" linear programming problem can be formed as follows:

$$\begin{aligned} &\min\left(\sum p_i + \sum q_i\right) \\ &\text{subject to} \\ &\boldsymbol{w}_c + \boldsymbol{W}\lambda + \boldsymbol{p} - \boldsymbol{q} = 0 \\ &\boldsymbol{p} \geq 0 \\ &\boldsymbol{q} \geq 0 \\ &\lambda \geq 0 \end{aligned} \tag{3.13}$$

This expanded system has a feasible solution for any given $\boldsymbol{\lambda}$ by suitable choices of $\boldsymbol{p}$ and $\boldsymbol{q}$. The minimum is zero if and only if the original system has a feasible solution.

3.8 Verification of Force Closure for 2-D Workpiece

The force closure for a four-contact frictionless fixturing configuration can be directly verified through matrix manipulation. Consider an object constrained by four point contacts as illustrated in Figure 3.7, the following equation can be derived from the equilibrium condition:

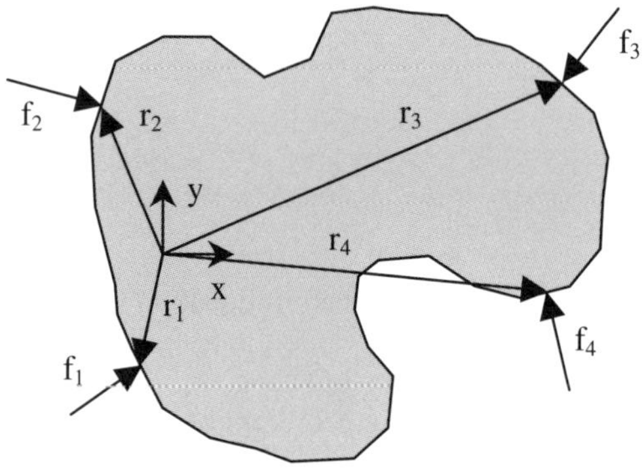

Figure 3.7 Four-contact frictionless fixturing

$$\begin{bmatrix} \cos\alpha_1 & \cos\alpha_2 & \cos\alpha_3 & \cos\alpha_4 \\ \sin\alpha_1 & \sin\alpha_2 & \sin\alpha_3 & \sin\alpha_4 \\ \boldsymbol{m}_1 & \boldsymbol{m}_2 & \boldsymbol{m}_3 & \boldsymbol{m}_4 \end{bmatrix} \begin{bmatrix} \lambda_1 \\ \lambda_2 \\ \lambda_3 \\ \lambda_4 \end{bmatrix} = 0 \tag{3.14}$$

where $\boldsymbol{m}_i = \boldsymbol{r}_{ix} \sin\alpha_i - \boldsymbol{r}_{iy} \cos\alpha_i$ and $(\cos\alpha_i, \ \sin\alpha_i)$ is the unit reaction force at the ith contact and α_i is the inner normal direction at the contacts with respect to the $+x$ axis.

$$\boldsymbol{M} = \begin{bmatrix} \cos\alpha_1 & \cos\alpha_2 & \cos\alpha_3 & \cos\alpha_4 \\ \sin\alpha_1 & \sin\alpha_2 & \sin\alpha_3 & \sin\alpha_4 \\ \boldsymbol{m}_1 & \boldsymbol{m}_2 & \boldsymbol{m}_3 & \boldsymbol{m}_4 \end{bmatrix}$$

The solution of the homogeneous system of linear equations (3.14) must lie within the null space of the linear transformation matrix $\boldsymbol{M}$ and can be expressed by scaled multiple as $[\lambda_1 \ \ \lambda_2 \ \ \lambda_3 \ \ \lambda_4]^T = \xi \ker(M)$ where $\xi \in \Re$. Force closure is achieved only when the solution satisfies that all $\lambda_i > 0$. If all the elements of the ker($\boldsymbol{M}$) are non-zero with same sign, strictly positive solution exists by appropriately choosing ξ. Therefore, force closure is provided by the fixture when all the components of the kernel of the matrix $\boldsymbol{M}$ are non-zero with same sign.

The fixturing case in Figure 3.1 is now examined using the present method. The positions of the contacts and the inner normal directions are listed in Table 3.1 and the matrix $\boldsymbol{M}$ is calculated as follows:

Table 3.1 Positions and normals of the contacts

1	$r_1=(15\sqrt{3},\ 15)$	$\alpha_1=300°$
2	$r_2=(10,\ 0)$	$\alpha_2=90°$
3	$r_3=(90,\ 0)$	$\alpha_3=90°$
4	$r_4=(100+\frac{1}{2}\boldsymbol{l},\ \frac{\sqrt{3}}{2}\boldsymbol{l}),\ 0\le \boldsymbol{l}\le 100$	$\alpha_4=150°$

$$\boldsymbol{M}=\begin{bmatrix} \frac{1}{2} & 0 & 0 & -\frac{\sqrt{3}}{2} \\ -\frac{\sqrt{3}}{2} & 1 & 1 & \frac{1}{2} \\ -30 & 10 & 90 & 50+\boldsymbol{l} \end{bmatrix}$$

The kernel of matrix $\boldsymbol{M}$ can be computed using the complete pivoting method (O'Neil, 1993). It is found that not all its components have the same sign and hence force closure is impossible to be achieved.

$$\ker(\boldsymbol{M})=\begin{bmatrix}\sqrt{3} & 1.1+0.0125\boldsymbol{l} & -(0.1+0.0125\boldsymbol{l}) & 1\end{bmatrix}^T$$

However, once the locating pin No.1 is shifted to the middle of the edge at $\boldsymbol{r}_1=(75,\ 25\sqrt{3})$, it is feasible to achieve force closure and hence the fixturing configuration becomes viable. The matrix $\boldsymbol{M}$ and its kernel are found to be:

$$\boldsymbol{M}=\begin{bmatrix} \frac{1}{2} & 0 & 0 & -\frac{\sqrt{3}}{2} \\ -\frac{\sqrt{3}}{2} & 1 & 1 & \frac{1}{2} \\ -50\sqrt{3} & 10 & 90 & 50+\boldsymbol{l} \end{bmatrix}$$

$$\ker(\boldsymbol{M})=\begin{bmatrix}\sqrt{3} & -\frac{1}{8}+\frac{\boldsymbol{l}}{80} & \frac{9}{8}-\frac{\boldsymbol{l}}{80} & 1\end{bmatrix}^T$$

It is clear from the kernel component values that when the clamp is placed at the range of $10 < l < 90$, force closure is always attained.

3.9 Criteria for Clamping Layout Reasoning

The convex hull approach is a general methodology for testing force closure. A method to inspect whether the origin lies inside a convex hull must be developed. Supposing a convex hull in d-dimensional space is bounded by s facets, then each facet can be represented by a hyper-plane described by its normal vector and an offset distance from the origin. The boundary facet is oriented correctly such that the distance from an interior point to the facet is negative. Let $\boldsymbol{A} = (\boldsymbol{a}_{ij})$ denote an $s \times d$ matrix of facet normals and $\boldsymbol{b}$ represent an s-component vector of facet offsets, any points inside the convex hull must satisfy:

$$\begin{bmatrix} a_{11} & a_{12} & \cdots & a_{1d} \\ a_{21} & a_{22} & \cdots & a_{2d} \\ \vdots & & & \vdots \\ a_{s1} & a_{s2} & \cdots & a_{sd} \end{bmatrix} \begin{bmatrix} x_1 \\ x_2 \\ \vdots \\ x_d \end{bmatrix} + \begin{bmatrix} b_1 \\ b_2 \\ \vdots \\ b_s \end{bmatrix} \le 0 \qquad (3.15)$$

or

$$\boldsymbol{Ax} + \boldsymbol{b} \le 0$$

Therefore, the convex hull includes the origin if and only if all the normal-offset values are non-positive:

$$\boldsymbol{b}_i \le 0 \qquad i = 1, 2, ..., s$$

The position of the origin can be successively categorised into the following states:

- If any offset value is positive, then the origin must lie outside the convex hull.
- If all the offset values are strictly negative, then the origin must lie strictly inside the convex hull, otherwise
- The origin lies on the boundary of the convex hull when one or more offset values are zero.

Consequently, the geometrical approach to exploring whether a fixturing configuration achieves the state of equilibrium can be mathematically summarised as follows:

- All $b_i < 0$, force closure.
- All $b_i \leq 0$ with one or more $b_i = 0$, marginal equilibrium. (3.16)
- Any $b_i > 0$, non-equilibrium.

An alternative method to determine the existence of an interior origin is to calculate the volume of the convex hull. If there is any change in volume when the origin is added to the bounding wrench set, the origin must be outside the convex hull, otherwise it is inside the convex hull. However, one limitation associated with this method is that it would reduce the computing efficiency as it needs to construct the convex hull twice and marginal equilibrium cannot be differentiated from force closure.

The criterion expressed in Equation (3.16) is a clear-cut binary metric: a set of contacts can, or cannot, guarantee force closure. A quantitative measure needs to be used in order to rank the goodness of a system of contacting wrenches, hopefully to generate a viable fixture more quickly. Possible useful quantifying heuristics include: the radius of the largest enclosed sphere centred at the origin, the volume of the convex hull, etc.

Definition For a convex hull containing the origin in $\mathfrak{R}^d$, an inside hyper-sphere with centre at the origin and radius $\boldsymbol{r} \geq 0$ is defined as the set of points $\boldsymbol{X} = \{\boldsymbol{x} \mid \|\boldsymbol{x}\| \leq \boldsymbol{r}\}$ such that every point of the set is in the convex hull.

In the case of marginal equilibrium, the inscribed hyper-sphere has zero radius as the origin lies on the boundary of the convex hull. This provides a useful heuristic to use the radius $\boldsymbol{r}_{\max}$ of the maximal inscribed hyper-sphere as the quantitative measure. There is an infinite number of inside hyper-spheres centred at the origin within the convex hull. The largest possible inscribed hyper-sphere is always the one with the shortest distance from the origin to the facets as its radius. Therefore, $\boldsymbol{r}_{\max}$ can be calculated using following expression:

$$r_{\max} = \min_{i \in [1,2,\ldots,s]} \left(\frac{|b_i|}{\sqrt{a_{i1}^2 + a_{i2}^2 + \cdots + a_{id}^2}} \right) \tag{3.17}$$

The conditions expressed in the propositions can be used to verify whether a particular configuration of fixturing elements is capable of achieving force closure and clamping equilibrium. In the case of clamping configuration optimisation, it can be used as criteria for reasoning out feasible clamping points on candidate clamping surfaces and determining an appropriate clamping sequence.

3.10 Automatic Generation of Clamping Points and Sequence

The problem of generating a viable clamping scheme can be stated as follows: "Given a dimensioned solid model of an arbitrarily-shaped workpiece and the locating positions, determine a feasible clamping configuration with optimal clamping points and positive clamping sequence such that the workpiece stability is always secured throughout the course of machining."

Synthesis of the viable fixturing function is to search for, among a large set of candidate fixturing regions, a finite number of fixturing points for both locators and clamps to constrain a workpiece. The common fixturing practice of the 3-2-1 locating principle can be employed to accomplish a locating configuration. Locating faces and positions may be selected based on some heuristic rules and physical requirements (Nee *et al.*, 1995). In this chapter, the fixturing points in question are restricted to clamping points as the locating points can be defined by the locating configuration synthesis method in previous chapters.

For a 3-D workpiece, the number of clamps may range from one to three. It is desirable that the clamps are positioned in such a way that there is no duplication of the restraining function. This means that clamps should not be placed on any locating plane. Parallel clamping or multi-clamping on the same surface should be avoided as far as possible. It also means the clamping directions should be oriented orthogonally as much as possible. This consideration allows only major surfaces to be taken into account for clamping reasoning and trivial surfaces to be discarded, resulting in quicker solutions.

The input to the clamping generation algorithm is as follows:

- Coefficient of static friction for each contacting point;
- Positions and orientations of locators, provided as a set of position vectors and normal vectors;
- Polygonal boundary of candidate clamping faces represented in terms of a list of ordered vertices for each face. In case of curved surfaces, an implicit relation $S(x, y, z) = 0$ describing the surface should be provided;
- A set of clamp access directions for all candidate clamping faces. If the clamping face is planar, then the clamp access direction is a normal vector

directed towards the workpiece. If the clamping face is a contoured surface, the inner normal vectors may be derived from the implicit relation.

The underlying strategy in the clamping point reasoning is a generate-and-test scheme by which invalid clamping points are pruned through checking all the candidate clamping points on a clamping face in a sequential point-by-point manner. The reasoning algorithm of clamping points and sequence can be formulated as the following steps:

1. Generate all the primitive and bounding wrenches for the locating contacts.

2. Generate a mesh over each clamping face, where the nodes on the mesh represent all the possible points that a clamp can exert force on the workpiece. If the candidate clamping face is a non-convex polygon or a spherical or cylindrical surface, it can be meshed through the triangulation procedure.

3. Generate the primitive and bounding wrenches for a candidate clamping point on clamping faces. Construct the convex hull of all the bounding wrenches using Quickhull algorithm and compute their facet offsets b_i.

4. Feasible clamping point is determined by checking for the negative facet offsets with $\boldsymbol{b}_i < 0 \ \forall \boldsymbol{i}$.

5. Compute the radius $\boldsymbol{r}_{\max}$ of the maximum inscribed hyper-sphere using Equation (3.17) for the feasible clamping point. Repeat Steps 3 through 5 until all the candidate clamping points are exhausted.

6. Sort all the feasible clamping regions constituted by the feasible clamping points on all clamping faces according to their area. Place the first clamp on a point with maximum $\boldsymbol{r}_{\max}$ from the largest clamping region. Check for existence of feasible solution to Equation (3.12) using linear programming technique in (3.13) for clamping equilibrium.

7. Place the second clamp, if any, on a point with maximum $\boldsymbol{r}_{\max}$ from the second largest clamping region, and so forth. Here the clamping sequence is inherently decided.

8. If no feasible clamping points are found, then the locating layout is recommended to be redesigned.

The direct correlation between the spatial geometrical reasoning for fixturing elements and the convex hull of bounding contacting wrenches enables the reasoning process to be mathematically established. It serves as a foundation for a theory of optimum clamping configuration for an automated fixture design system.

Section 3.12 provides case studies to demonstrate the methodology for determining optimal clamping points and sequence.

3.10.1 Consideration of Physical Constraints

The reasoning procedure demonstrates the automatic generation of viable clamping positions, which may occur at arbitrary locations in space depending on the mesh generated. In the case of modular fixtures, there are structural constraints on the fixturing layout. The locations of fixturing contacts are restricted by the lattice units of the modular fixture kit. The number of possible clamping points may be substantially reduced by the physical constraints. The structural constraint problem can be solved by slight repositioning of the clamps such that they are on the nearest intersections of the feasible clamping regions and physical constraints.

Another kind of physical constraint is the safety concern of cutting tool interference. To prevent the obstruction of the clamps with the cutting tool as it makes a pass, the clamps should not be placed within a specified minimum distance from the machining paths. The interference problem can be resolved by discarding all the clamping points within the minimum distance along the tool paths.

3.11 Algorithm Implementation

The clamp reasoning algorithm is developed and implemented on an HP Apollo workstation. The formulation is carried out with the assistance of a symbolic algebra package Maple V (Abell and Braselton, 1994). The reasoning algorithm yields a set of optimal clamping points with respect to the proposed metric. Nevertheless, other optimal metrics may be incorporated to produce specific configurations. For example, minimal clamping intensity could be used to rank the feasible clamping points. In addition to the best clamping set, all the feasible clamping points are available to a tool designer who may select an appropriate set of clamping points from the clamping pool so that ill-conceived configurations could be avoided.

The variables involved in the algorithm are the nodal position coordinates $[\boldsymbol{x}, \boldsymbol{y}, z]^T$ associated with a candidate clamping point. The nature of the clamping analysis algorithm is hence essentially an enumeration procedure in which a finite number of possible clamping points are examined for the clamping validity. The number of possible clamping points to be considered is at the discretion of the user through the specification of mesh density. In consideration of the existence of physical constraints, it is not essential for the algorithm to completely check the entire area on a clamping face. In practice, it is sufficient to produce a set of optimal clamping points from the discrete nodes of the mesh only.

3.12 Case Studies

3.12.1 Planar Fixturing Problems

The developed methodology is first illustrated by a 2-D workpiece (Figure 3.8). A rectangular workpiece is located by three round locating pins. Assuming the coefficient of static friction to be $\mu = 0.2$ and the inner normal vector can be expressed as $[\cos(\alpha_i), \sin(\alpha_i)]^T$ at contact point C_i, then the bounding wrenches $\overline{w}_{11}, \overline{w}_{12}, \cdots, \overline{w}_{42}$ associated with the contacting points are calculated from the generating vectors $f_{11}, f_{12}, \cdots f_{42}$ of the friction sectors as follows:

$$\overline{w}_{i1} = [\cos(\alpha_i - \theta), \quad \sin(\alpha_i - \theta), \quad x_i \sin(\alpha_i - \theta) - y_i \cos(\alpha_i - \theta)]^T$$
$$\overline{w}_{i2} = [\cos(\alpha_i + \theta), \quad \sin(\alpha_i + \theta), \quad x_i \sin(\alpha_i + \theta) - y_i \cos(\alpha_i + \theta)]^T$$

where $\theta = \tan^{-1}(\mu)$ and (x_i, y_i) is the position coordinates of C_i. The wrenches are computed and tabulated in Table 3.2.

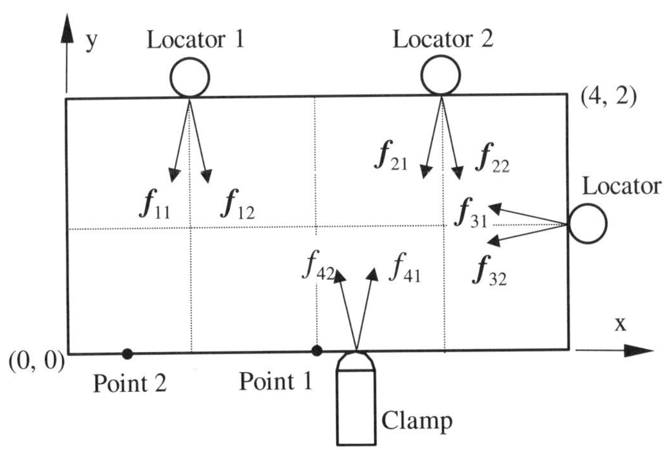

Figure 3.8 A 2-D clamping point reasoning problem

Table 3.2 Bounding wrenches for a planar fixturing problem

Wrench	Clamped at *y*-direction	Clamped at *x*-direction
$\overline{w}_{11}$	$[-0.1961, -0.9805, -0.5883]^T$	$[-0.1961, -0.9805, -0.5883]^T$
$\overline{w}_{12}$	$[0.1961, -0.9805, -1.3728]^T$	$[0.1961, -0.9805, -1.3728]^T$
$\overline{w}_{21}$	$[-0.1961, -0.9805, -2.5495]^T$	$[-0.1961, -0.9805, -2.5495]^T$
$\overline{w}_{22}$	$[0.1961, -0.9805, -3.3339]^T$	$[0.1961, -0.9805, -3.3339]^T$
$\overline{w}_{31}$	$[-0.9805, 0.1961, 1.7650]^T$	$[-0.9805, 0.1961, 1.7650]^T$
$\overline{w}_{32}$	$[-0.9805, -0.1961, 0.1961]^T$	$[-0.9805, -0.1961, 0.1961]^T$
$\overline{w}_{41}$	$[0.1961, 0.9805, 0.9805x]^T$	$[0.9805, -0.1961, -0.9805y]^T$
$\overline{w}_{42}$	$[-0.1961, 0.9805, 0.9805x]^T$	$[0.9805, 0.1961, -0.9805y]^T$

3.12.1.1 Verification of Force Closure

As shown in Figure 3.8, if one clamp is placed at point 1, then force closure can be achieved. This can be clearly illustrated by a graphical representation of the convex hull in Figure 3.9 where the origin is strictly inside the convex hull. On the other hand, if one clamp is positioned at point 2 in Figure 3.8, the clamping point is found invalid because equilibrium cannot be achieved here as the origin lies outside the convex hull as shown in Figure 3.10. Using the developed method, the existence of an interior origin is easily recognised by examining the facet normal offsets as tabulated in Table 3.3 where point 2 is readily identified as an invalid clamping point for its positive offset values.

3.12.1.2 Best Clamping Points and Positive Clamping Sequence

In this fixturing problem, each restraint is in frictional hard point contact with the workpiece. Using the clamping points reasoning procedure, the feasible clamping regions are generated and then highlighted by bold segments in Figure 3.11(a). It is found that one clamp *alone* is sufficient to achieve force closure. From the point of view of force closure, this clamp can be positioned either on line segment *ab* or on line segment *cd* (in bold line). However, by considering the clamping equilibrium condition, only a portion of the line segment *ab* (line segment *ef*) and a point on *cd* (point *g*) can be used as valid clamping region (Figure 3.11(a)). Point *g* is strongly discouraged to be considered as a clamping point as in practice the clamp always has positioning error, which may cause the location of the workpiece to be disturbed. Finally, the clamping point is chosen to be at point 1 where $\boldsymbol{r}_{\mathbf{max}}$ is the largest of all the feasible points on the side as shown in Figure 3.12.

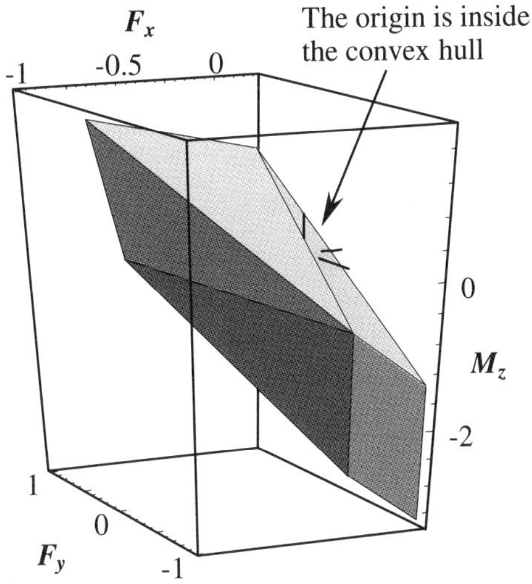

Figure 3.9 Convex hull of bounding wrenches containing the origin when the clamp is placed at point 1

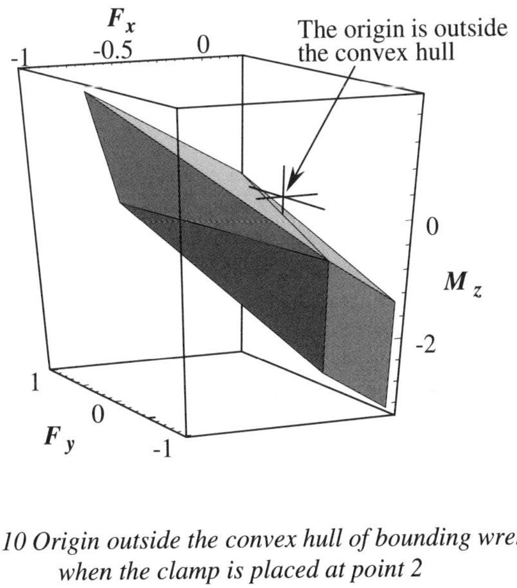

Figure 3.10 Origin outside the convex hull of bounding wrenches when the clamp is placed at point 2

Table 3.3 Facet normal offsets of convex hulls

Facet	Normal offsets when clamped at point 1	Normal offsets when clamped at point 2
b_1	-0.1961	-0.1961
b_2	-0.2315	-0.5271
b_3	-0.2383	-0.6488
b_4	-0.4838	-0.6502
b_5	-1.6647	-1.0920
b_6	-0.5462	-0.5462
b_7	-0.2443	**0.0201**
b_8	-0.3526	**0.0229**
b_9	-0.7692	-0.7692
b_{10}	-0.8320	-0.8320
b_{11}	-0.9805	-0.9805
	Feasible clamping	Invalid clamping

In practice, more clamps than force closure requirement may be employed in order to avoid excessive clamping force. In the case of using two clamps, the position of a second clamp needs to be decided. Remember that the clamping equilibrium requirement is no longer imposed on the subsequent clamp; it is chosen to be placed at point 2 where $\boldsymbol{r}_{\mathbf{max}}$ is the largest of all the feasible points on the side as shown in Figure 3.13. Here the clamping sequence is inherently determined. The final fixturing configuration is shown in Figure 3.11(b) where the second clamp depicted in hidden line is optional and the positive clamping sequence is to place first clamp on point 1, followed by second clamp on point 2.

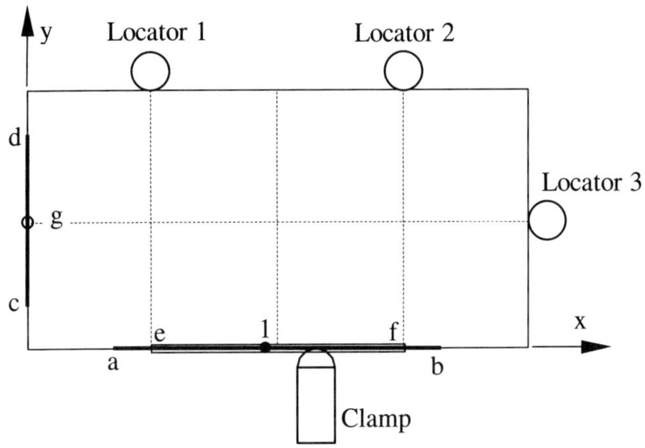

(a) Feasible clamping region (bold segments)

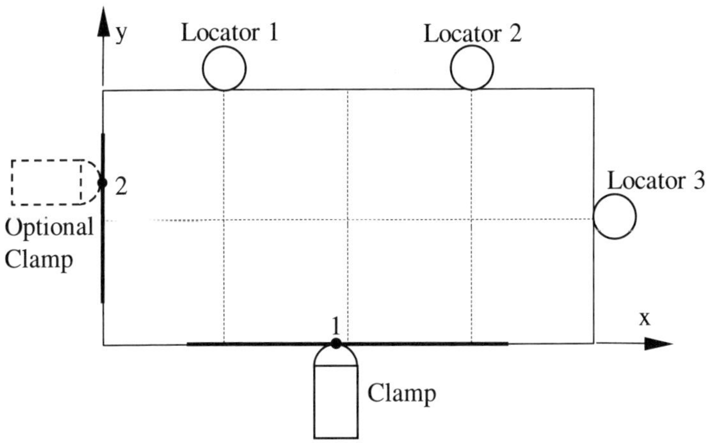

(b) Final fixturing layout

Figure 3.11 Generation of a viable fixturing configuration for a 2D workpiece

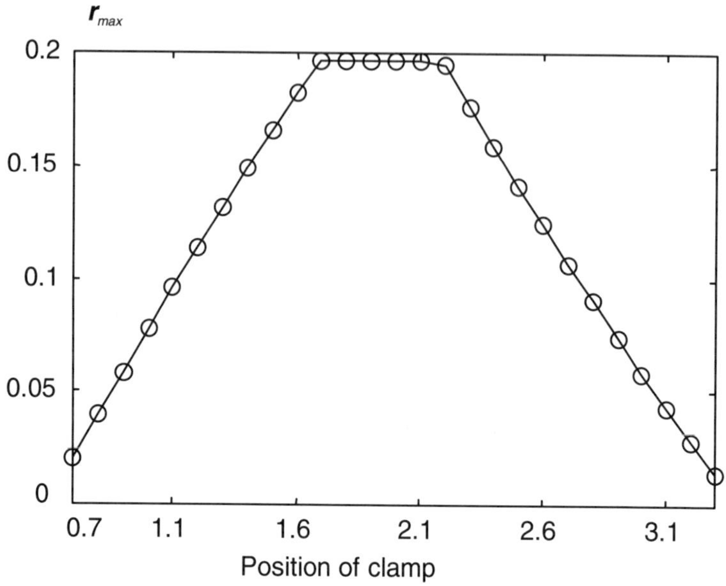

Figure 3.12 The radius of maximal inscribed hyper-sphere vs. the position of clamp along the x-direction

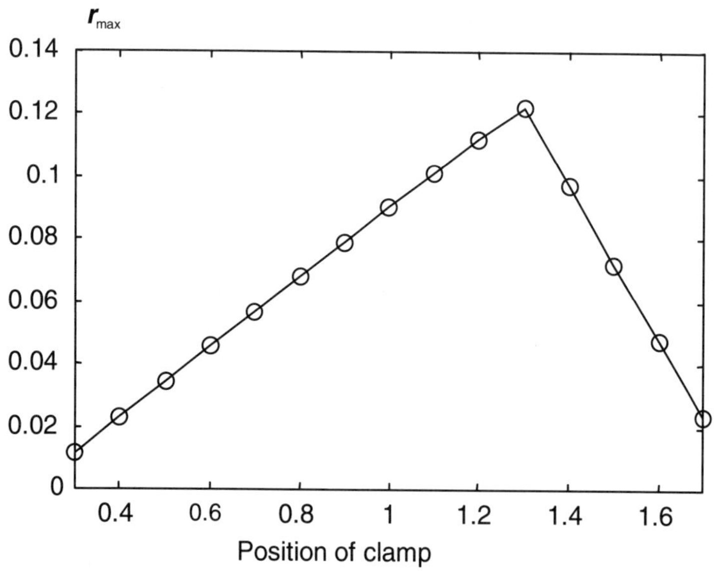

Figure 3.13 The radius of maximal inscribed hyper-sphere vs. the position of clamp along the y-direction

3.12.2 Spatial Fixturing Problems

3.12.2.1 Prismatic Workpiece

A prismatic workpiece is next explored for fixturing configuration as shown in Figure 3.14. The block workpiece with overall size of 220(L) × 120(W) × 100(H) mm is located by six locator pins according to the 3-2-1 locating principle. All the restraints are assumed frictional and each friction cone is approximated by a 4-sided polyhedral cone. Consequently, the number of the six-dimensional bounding wrenches associated with the six locating contacts and one clamping contact sums to 28. There are three surfaces available for clamping: top, side and end faces. Two clamping schemes are possible: three clamps (one downward and two side clamps) applied to all the clamping faces or two side clamps only if the machining operation on the top face prevents placing any clamp on it.

The candidate top, side and end faces are first discretised into 44 × 24, 44 × 20 and 24 × 20 meshes respectively as shown in Figures 3.15, 3.16 and 3.17 respectively. All the mesh nodes are examined for clamping feasibility by the developed algorithm. A division area bounded by valid clamping points is referred to as a feasible clamping region. The feasible regions for top, side and end faces are shown as shaded areas in Figures 3.15, 3.16 and 3.17 respectively. It should be pointed out that the feasible areas would become slightly different if a 6- or 8-sided approximation polyhedron cone is used. Nevertheless, the 4-sided polyhedron approximation is sufficient to produce an adequate feasible clamping area.

The best clamping points are decided to be at (70, 60, 100) for top face, at (80, 0, 60) for side face and at (0, 45, 50) for end face with respect to the $\boldsymbol{r}_{\max}$ measure. This is accomplished by searching for the maximum $\boldsymbol{r}_{\max}$ value of all the valid clamping points on the corresponding clamping faces. At those points, $\boldsymbol{r}_{\max}$ reaches the peak values of 0.1105, 0.0638 and 0.0491 respectively. The feasible region area on the top face is the largest while that on the end face is the smallest. Consequently, the first clamp should be applied on the top face, followed by a second clamp placed on the side face and a third clamp on the end face. The first clamp is then checked for clamping equilibrium condition and is found valid.

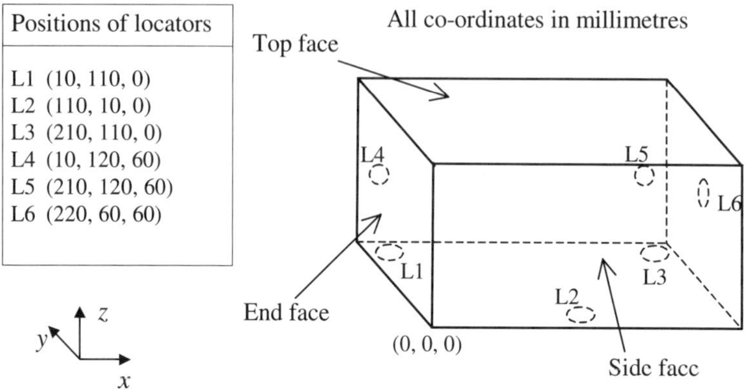

Figure 3.14 A typical example of 3-D clamping reasoning problem

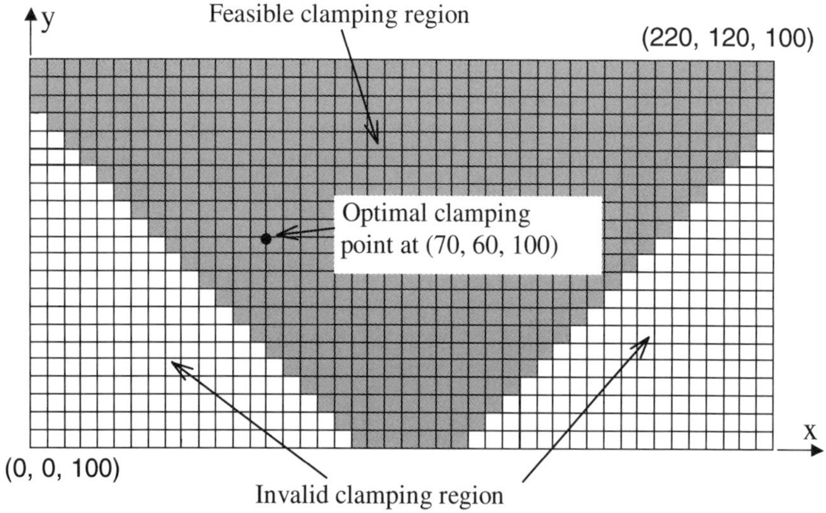

Figure 3.15 Surface mesh on top face and the generated feasible clamping region

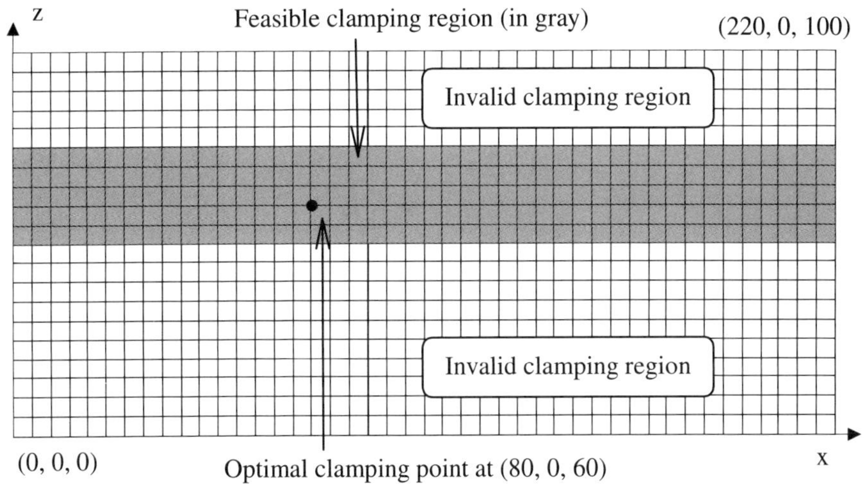

Figure 3.16 Surface mesh on side face and the generated feasible clamping region

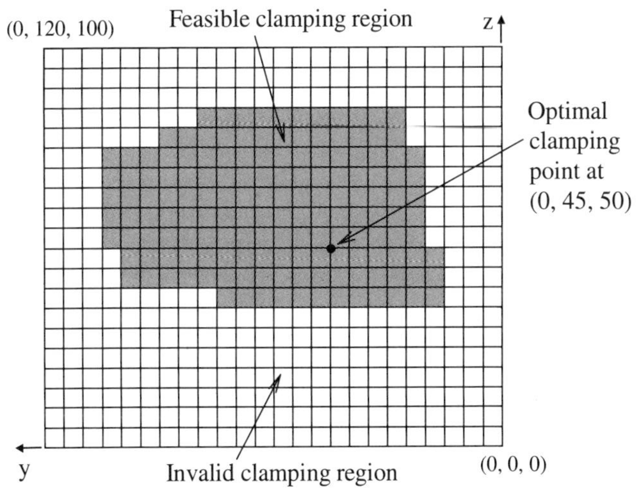

Figure 3.17 Surface mesh on end face and the generated feasible clamping region

3.12.2.2 Engineering Workpiece

The next case study is a more complex workpiece as shown in Figure 3.18. Note that the primary locators *L1*, *L2* and *L3* are not on the same plane as there is a shoulder feature on the underside. The positions of all the locators are listed in Figure 3.18. An arcuate surface feature is to be machined using a CNC milling machine.

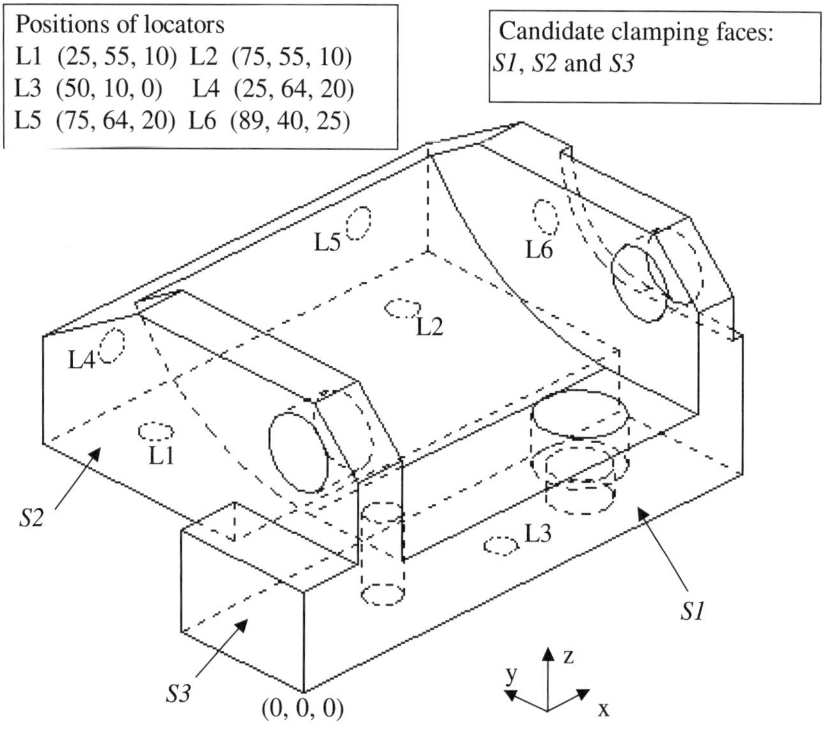

Figure 3.18 A case study of clamping configuration generation

In this operation, there are three planar surfaces available for clamping: *S1*, *S2* and *S3*. Each candidate clamping surface can be represented by its polygonal boundary. Note that not all the polygons are convex. The clamping surfaces are then discretised into meshes through a triangulation procedure. Figure 3.19(a) shows the feasible clamping region in crosshatch for the surface *S1* generated by the developed algorithm. The optimal clamping point is determined to be at (59, 0, 13) where $\boldsymbol{r}_{max}$ reaches the maximum value of 0.0235 as illustrated in a contour plot in Figure 3.19(b). The feasible clamping region on the surface *S2* is next generated as illustrated in Figure 3.20(a). The optimal clamping point is found to be at (11, 36, 20) with the maximum $\boldsymbol{r}_{max}$ value of 0.0417 (Figure 3.20(b)). However, the surface *S3* is found to be an invalid clamping face as none of the clamping points on it can achieve force closure. It is hence disqualified as a feasible clamping face.

Therefore, the optimal clamping configuration consists of two side clamps on surfaces *S1* and *S2* located at the generated optimal clamping points. The clamping sequence is determined to be the first clamp on *S2* followed by the second clamp on *S1* according to area size of individual feasible clamping regions. The first clamp is also found viable for clamping equilibrium.

3.13 Summary

The clamping layout synthesis in this chapter is summarised as follows:

- A computational geometry approach to the verification of force closure of a fixturing configuration with the presence of friction is first derived. A supplementary condition of clamping equilibrium is then proposed.

- Constructive criterion for reasoning viable clamping layout is proposed so that fixturing configurations can be verified and rated by a quantitative metric.

- Reasoning procedure for generating optimal clamping points and positive clamping sequences is developed and validated.

This chapter presents a geometrical reasoning methodology for determining optimal clamping points and clamping sequence. The method provides a simple and yet effective means for determining a set of optimal clamping points from the candidate clamping layouts with respect to the metric of force closure. Clamping sequence is inherently dealt with in this approach. The developed approach has the ability to tackle complex arbitrarily-shaped workpieces. The present methodology is robust enough to handle various fixturing problems. Another advantage of the method is its ease in formulating and achieving solutions. The algorithm has been verified using several different case studies.

The approach is useful for the optimisation of fixturing schemes, which traditionally depends mainly on the tool designer's expertise and experience. When working in conjunction with a clamping force optimisation system, the proposed analysis contributes to fixture design improvement before the development of fixture prototype.

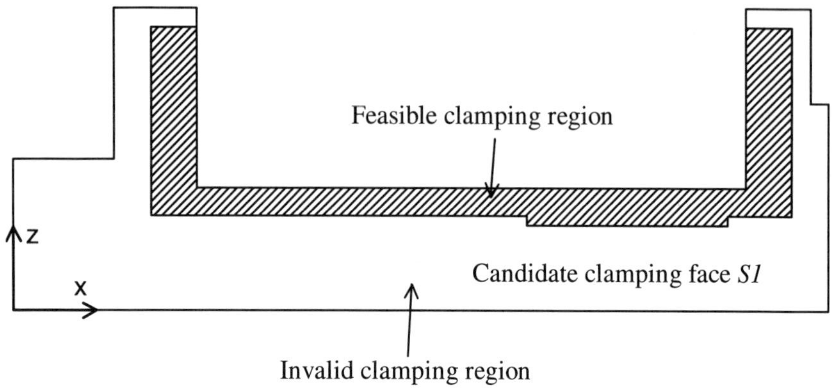

(a)

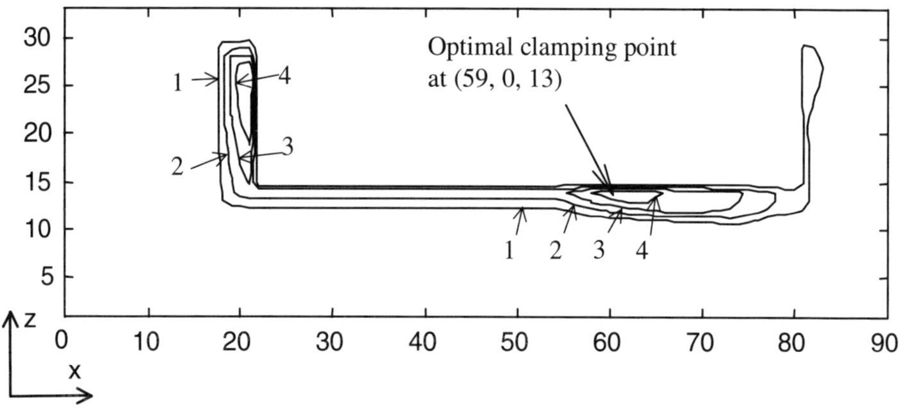

(b)

Figure 3.19 (a) Feasible clamping region generated on face S1 and (b) contour plot of $\mathbf{r}_{max}$ distribution over the feasible clamping region.

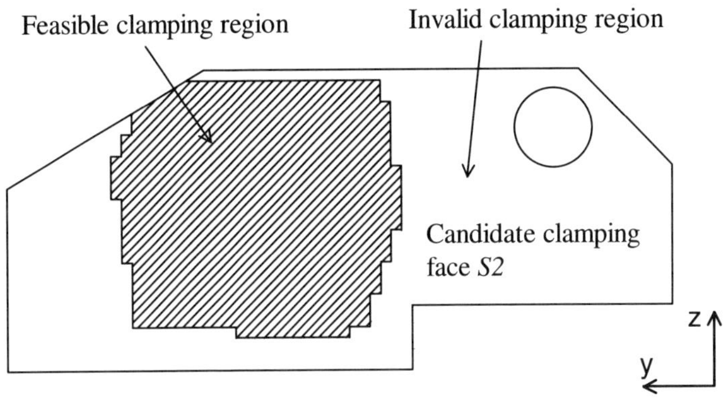

(a)

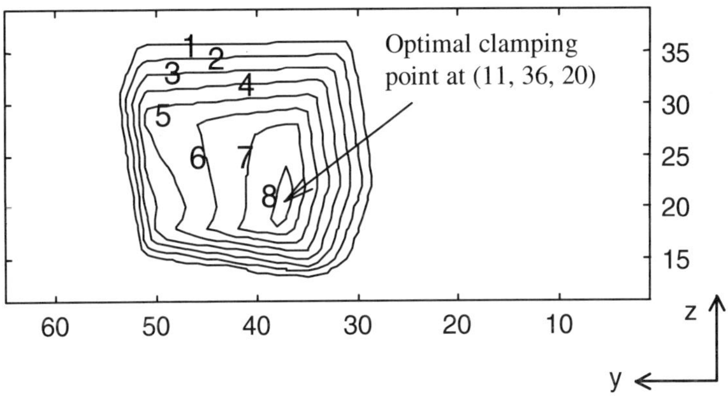

(b)

Figure 3.20 (a) Feasible clamping region generated on face S2 and (b) contour plot of $\mathbf{r}_{max}$ distribution over the feasible clamping region

3.14 Bibliography

Abell, M.L. and J.P. Braselton. *The Maple V Handbook*. Boston: AP Professional. 1994.

Asada, H. and A.B. By. Kinematic Analysis of Workpart Fixturing for Flexible Assembly with Automatically Reconfigurable Fixtures. *IEEE Journal of Robotics and Automation*, 1(2), pp. 86-94. 1985.

Barber, C.B., D.P. Dobkin and H.T. Huhdanpaa. The Quickhull Algorithm for Convex Hulls. *ACM Transactions on Mathematical Software*, 22, pp. 469-483. 1996.

Cecil, J. Clamping Design Approach for Automated Fixture Design. *Int. J. Adv. Mfg. Technol.*, Vol.18, pp.784-789. 2001.

Chou, Y.-C., V. Chandru and M.M. Barash. A Mathematical Approach to Automatic Configuration of Machining Fixtures: Analysis and Synthesis. *Journal of Engineering for Industry*, 111(4), pp.579-584. 1989.

Chvatal, V. *Linear Programming*. W.H. Freeman and Company. 1983.

DeMeter, E.C. Restraint Analysis of Fixtures which Rely on Surface Contact. *Journal of Engineering for Industry*, 116(2), pp. 207-215. 1994.

Gologlu, C. Machine Capability and Fixturing Constraints-imposed Automatic Machining Set-ups Generation. *Journal of Materials Processing Technology*, Vol.148, No.1, pp.83-92. 2004.

Hargrove S.K. and A. Kusiak. Computer-aided Fixture Design: a Review. *International Journal of Production Research*, 32(4), pp.733-753. 1994.

Hazen, F.B. and P. Wright. Autonomous Fixture Loading by a Machine Tool. *Proc. of ASME Symp. on Manuf. Sys.*, pp. 105-111. 1988.

Hazen, F.B. and P. Wright. Workholding Automation: Innovations in Analysis, Design and Planning. *Manufacturing Review*, 3(4), pp.224-236. 1990.

Hirai, S. and H. Asada. Kinematics and Statics of Manipulation Using the Theory of Polyhedral Convex Cones. *Int. J. Robotics Research*, 12(2), pp.434-447. 1993.

Hong, D. and C.I. Lin. Development of an Automated Flexible Fixture for Planar Objects. *Robotics and Computer-Integrated Manufacturing*, Vol.14, No.3, pp.173-183. 1998.

Kerr, J. and B. Roth. Analysis of Multi-fingered Hands. *Int. J. Robotics Research*, 4(4), pp.3-17. 1986.

Kline, W.A. *The Prediction of Cutting Forces and Surface Accuracy for the End Milling Process*. PhD Thesis, University of Illinois at Urbana-Champaign. 1982.

Krishnakumar, K. and S.N. Melkote. Machining Fixture Layout Optimization Using the Genetic Algorithm. *International Journal of Machine Tools and Manufacture*, Vol.40, No.4, pp.579-598. 2000.

Lakshminarayana, K. *Mechanics of Form Closure*. Technical report 78-DET-32, ASME. 1978.

Lee, S.H. and M.R. Cutkosky. Fixturing Planning with Friction. *Journal of Engineering for Industry*, 113, pp. 320-327. 1991.

Liao, Y.G. A Generic Algorithm-based Fixture Locating Positions and Clamping Schemes Optimization. *Proc. Instn. Mech. Engrs., Part B: J. Engineering Manufacture*, Vol.217, pp.1075-1083. 2003.

Mishra, B. and N. Silver. Some Discussion of Static Gripping and its Stability. *IEEE Sys. Man Cybernet.*, 19(4), pp. 783-796. 1989.

Mishra, B., J.T. Schwartz and M. Sharir. On the Existence and Synthesis of Multifinger Positive Grips. *Algorithmica.*, 2(4), pp.541-558. 1988.

Mittal, R.O., P.H. Cohen and B.J. Gilmore. Dynamic Modelling of the Fixture-Workpiece System. *Robotics and Computer-integrated Manufacturing*, 8(4), pp.201-217. 1991.

Nee, A.Y.C., K. Whybrew and A. Senthil kumar. *Advanced Fixture Design for FMS*. London: Springer-Verlag. 1995.

Nguyen, V.D. Constructing Force-closure Grasps. *Int. J. of Robotics Research*, 7(3), pp.3-16. 1988.

Nnaji, B.O., S. Alladin and P. Lyu. A Framework for a Rule-base Expert Fixturing System for Face Milling Planar Surface on a CAD System Using Flexible Fixtures. *Journal of Manufacturing System*, 7(3), pp.193-207. 1988.

O'Neil, P.V. *Advanced Engineering Mathematics*. Boston: PWS Publishing Co. 1993.

O'Rourke, J. *Computational Geometry in C*. Cambridge University Press. 1994.

Pelinescu, D.M. and Michael Wang. Multi-objective Optimal Fixture Layout Design. *Robotics and Computer-Integrated Manufacturing*, Vol.18, Issues 5-6, pp.365-372. 2002.

Ponce, J., S. Sullivan, J.D. Boissonnat and J.P. Merlet. On Characterising and Computing Three- and Four-Finger Force-Closure Grasps of Polyhedral Objects. *Proc. IEEE Int. Conf. on Robotics and Automation*, pp.821-827. 1993.

Salisbury, J.K. and B. Roth. Kinematics and Force Analysis of Articulated Hands. *J. Mechanisms Transmissions Automation Design*, 105, pp.33-41. 1982.

Trappey, A.J.C. and C.R. Liu. An Automatic Workholding Verification System. *Robotics and Computer-integrated Manufacturing*, 9(4/5), pp.321-326. 1992.

Wu, Y., Y. Rong, W. Ma and S.R. LeClair. Automated Modular Fixture Planning: Accuracy, Clamping, and Accessibility Analyses. *Robotics and Computer-Integrated Manufacturing*, Vol.14, No.1, pp.17-26. 1998.

Chapter 4

Optimisation of Dynamic Clamping Forces for a Fixture

4.1 Introduction

When a workpiece is properly located by a set of locators within a fixture, sufficient clamping force must be applied to hold it firmly against the locating surfaces and to resist the cutting forces acting upon it. It is important that the deterministic location of the workpiece should always be secured throughout the machining operation. Both over-clamping and under-clamping are detrimental to the accuracy of the workpiece. It is best to use minimal clamping forces that are just enough to counterbalance the external load during machining.

Therefore, it is always an essential concern in fixture design to compute or estimate the required clamping magnitude for a given fixture configuration. This chapter first introduces a direct computation approach for a 2-D fixture, by which the minimal clamping force is directly derived. A generic mathematical model for clamping intensity generation is then presented, based on a proposed maximum friction capacity criterion. Frictional force is included in the model and completely represented by a proposed friction model. Dynamic machining condition is also considered in the prediction of the minimal clamping forces. Case studies are presented to demonstrate the effectiveness of the developed methodology.

4.2 Direct Computation Approach

4.2.1 Minimal Clamping Force for Frictionless Fixturing

Given a 2-D fixturing configuration, a direct computation approach is developed for obtaining the minimal clamping intensity required to resist an external load. Supposing the contacts between the workpiece and the fixturing elements are frictionless, all the contact forces are thus perpendicular to the edge in contact with the restraints, directing into the workpiece. Four restraints are sufficient for this particular case to totally constrain a workpiece: three locators and one clamp (Figure 4.1).

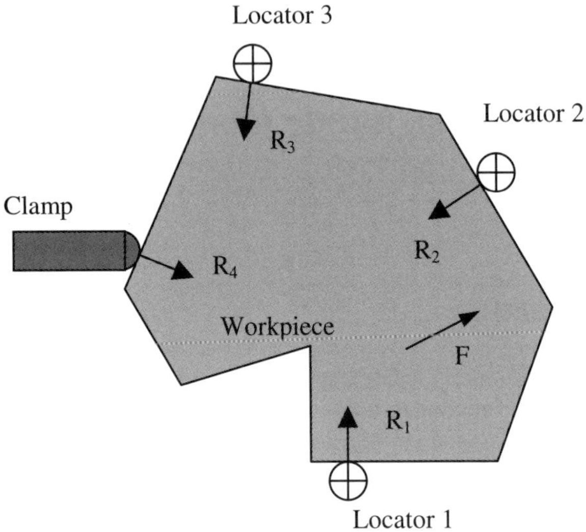

Figure 4.1 Frictionless 4-contact fixture

Reaction forces are induced by the external force $\boldsymbol{F}$ and clamping force $\boldsymbol{R}_4$. Their magnitudes can be deduced from static equilibrium condition as follows:

$$\begin{bmatrix} \boldsymbol{R}_1 \\ \boldsymbol{R}_2 \\ \boldsymbol{R}_3 \end{bmatrix} = \begin{pmatrix} \cos\alpha_1 & \cos\alpha_2 & \cos\alpha_3 \\ \sin\alpha_1 & \sin\alpha_2 & \sin\alpha_3 \\ \boldsymbol{m}_1 & \boldsymbol{m}_2 & \boldsymbol{m}_3 \end{pmatrix}^{-1} \begin{bmatrix} -\cos\alpha_4 \boldsymbol{R}_4 - \boldsymbol{F}_x \\ -\sin\alpha_4 \boldsymbol{R}_4 - \boldsymbol{F}_y \\ -\boldsymbol{m}_4 \boldsymbol{R}_4 + \boldsymbol{F}_x \boldsymbol{y}_c - \boldsymbol{F}_y \boldsymbol{x}_c \end{bmatrix} \tag{4.1}$$

where

$\boldsymbol{R}_i$ is the reaction force on the ith locator, i = 1, 2, 3;

$\boldsymbol{R}_4$ is the clamping force exerted by the clamp to the workpiece;

$[\boldsymbol{F}_x \quad \boldsymbol{F}_y]^T$ is the external load acting on the workpiece;

$\boldsymbol{m}_i = \boldsymbol{r}_{ix} \sin\alpha_i - \boldsymbol{r}_{iy} \cos\alpha_i$, i = 1, 2, 3, 4;

$(\cos\alpha_i, \ \sin\alpha_i)$ is the unit reaction force at the ith restraint, where α_i is the inner normal direction at the contacts with respect to +x axis;

$(\boldsymbol{x}_i, \ \boldsymbol{y}_i)$ is the location of the ith restraint's contact point;

$(\boldsymbol{x}_c, \ \boldsymbol{y}_c)$ is the application point of external load.

The first two equations ($\boldsymbol{R}_1$, $\boldsymbol{R}_2$) represent that the translational force components cancel each other. The last equation ($\boldsymbol{R}_3$) depicts that the torque induced by the external load is also balanced. Because the solution should be non-negative, the minimal clamping force can be directly found by incrementally increasing the $\boldsymbol{R}_4$ value from zero until all the $\boldsymbol{R}_1$, $\boldsymbol{R}_2$, $\boldsymbol{R}_3$ are greater than zero or a positive lower bound.

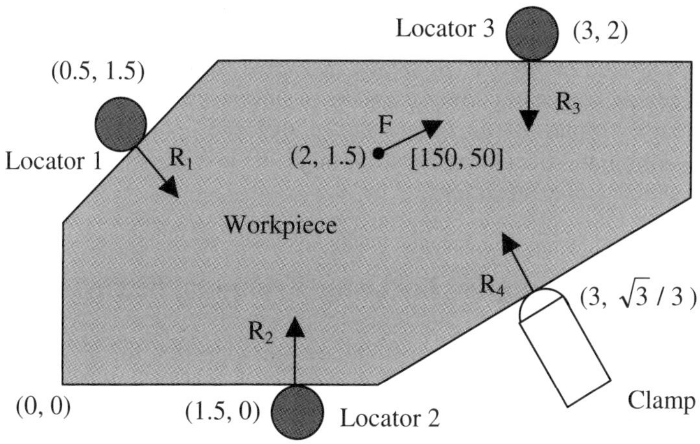

Figure 4.2 An example of a 2-D fixture

4.2.2 An Illustrating Example

The following is an example of the generation of minimal clamping force. A planar workpiece is located with three pins and secured by one clamp as shown in Figure 4.2. Assuming an external load $[\boldsymbol{F}_x, \boldsymbol{F}_y]$ = [150N, 50N] is acting on the workpiece, Equation (4.1) now becomes:

$$\begin{bmatrix} \boldsymbol{R}_1 \\ \boldsymbol{R}_2 \\ \boldsymbol{R}_3 \end{bmatrix} = \begin{pmatrix} 1/\sqrt{2} & 0 & 0 \\ -1/\sqrt{2} & 1 & -1 \\ -\sqrt{2} & 1.5 & -3 \end{pmatrix}^{-1} \begin{bmatrix} 0.5\boldsymbol{R}_4 - 150 \\ -0.5\sqrt{3}\boldsymbol{R}_4 - 50 \\ -5\sqrt{3}/3\boldsymbol{R}_4 + 125 \end{bmatrix}$$

Solving for $\boldsymbol{R}_1$, $\boldsymbol{R}_2$ and $\boldsymbol{R}_3$ and then setting them greater than zero, the following result is obtained:

$\boldsymbol{R}_4' \geq 300$

$\boldsymbol{R}_4'' \geq 538.87$

$\boldsymbol{R}_4''' \geq 93.44$

The minimal clamping intensity is thus determined as follows:

$$\boldsymbol{R}_4 = \max(\ \boldsymbol{R}_4',\ \boldsymbol{R}_4'',\ \boldsymbol{R}_4'''\ ) = 538.87\,\text{N}$$

If $\boldsymbol{R}_1$ or $\boldsymbol{R}_3$ is designated as the clamp, the required clamping force would be 168.90N or 397.24N respectively. If $\boldsymbol{R}_2$ is used as the clamp, substantial clamping force is not needed as the load is totally cancelled by the reaction forces on the locators. The importance of correct clamping position is clearly demonstrated in this case. From this result, it can be concluded that it is possible that an optimal clamping configuration could be found through minimising the required clamping force to counteract an external load.

4.2.3 Minimal Clamping Force for Frictional Fixturing

In general, three frictional contact points are sufficient for the construction of a planar workholding scheme with force closure. In practice, however, fixturing schemes involving four frictional contact points are commonly used as three locators are usually required to provide a deterministic location for a 2-D workpiece (Figure 4.3). Approach to the direct computation of the clamping force is derived in such a way that the external load imposed on the workpiece will be counterbalanced by a minimal clamping intensity.

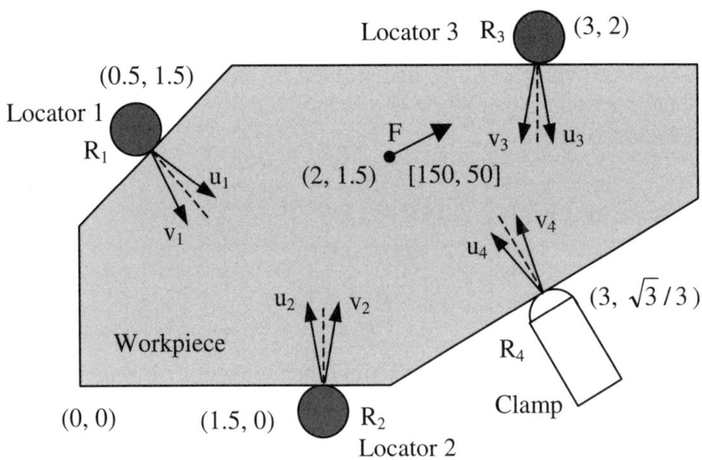

Figure 4.3 Generation of clamping intensity for a 2-D fixture

Contact force f_i at i^{th} contact point can be expressed as a non-negative linear combination of generating vectors $\boldsymbol{u}_i$ and $\boldsymbol{v}_i$ bounding the friction sector as shown in Figure 4.3:

$$\boldsymbol{f}_i = \boldsymbol{a}_i \boldsymbol{u}_i + \boldsymbol{b}_i \boldsymbol{v}_i \quad \text{with } \boldsymbol{a}_i \geq 0 \text{ and } \boldsymbol{b}_i \geq 0, \quad \text{i} = 1, 2, 3, 4 \tag{4.2}$$

Supposing the external forces $[\boldsymbol{F}_x, \boldsymbol{F}_y]$ imparted at $(\boldsymbol{x}_c, \boldsymbol{y}_c)$ and the clamp being the 4th restraint, the following system of equations can be derived from the static equilibrium condition:

$$\begin{pmatrix} \boldsymbol{u}_{1x} & \boldsymbol{v}_{1x} & \vdots & \boldsymbol{u}_{2x} & \boldsymbol{v}_{2x} & \vdots & \boldsymbol{u}_{3x} & \boldsymbol{v}_{3y} \\ \boldsymbol{u}_{1y} & \boldsymbol{v}_{1y} & \vdots & \boldsymbol{u}_{2y} & \boldsymbol{v}_{2y} & \vdots & \boldsymbol{u}_{3y} & \boldsymbol{v}_{3y} \\ \boldsymbol{m}_{u1} & \boldsymbol{m}_{v1} & \vdots & \boldsymbol{m}_{u2} & \boldsymbol{m}_{v2} & \vdots & \boldsymbol{m}_{u3} & \boldsymbol{m}_{v3} \end{pmatrix} \begin{bmatrix} \boldsymbol{a}_1 \\ \boldsymbol{b}_1 \\ \cdots \\ \boldsymbol{a}_2 \\ \boldsymbol{b}_2 \\ \cdots \\ \boldsymbol{a}_3 \\ \boldsymbol{b}_3 \end{bmatrix} = \tag{4.3}$$

$$= -\begin{pmatrix} \boldsymbol{u}_{4x} & \boldsymbol{v}_{4x} \\ \boldsymbol{u}_{4y} & \boldsymbol{v}_{4y} \\ \boldsymbol{m}_{u4} & \boldsymbol{m}_{v4} \end{pmatrix} \begin{bmatrix} \boldsymbol{a}_4 \\ \boldsymbol{b}_4 \end{bmatrix} - \begin{bmatrix} \boldsymbol{F}_x \\ \boldsymbol{F}_y \\ \boldsymbol{x}_c \boldsymbol{F}_y - \boldsymbol{y}_c \boldsymbol{F}_x \end{bmatrix}$$

where

$(\boldsymbol{u}_{ix} \quad \boldsymbol{u}_{iy}) = (\cos(\alpha_i + \theta) \quad \sin(\alpha_i + \theta))$ is the unit vector $\boldsymbol{u}_i$;

$(\boldsymbol{v}_{ix} \quad \boldsymbol{v}_{iy}) = (\cos(\alpha_i - \theta) \quad \sin(\alpha_i - \theta))$ is the unit vector $\boldsymbol{v}_i$;

$\boldsymbol{m}_{ui} = \boldsymbol{x}_i \sin(\alpha_i + \theta) - \boldsymbol{y}_i \cos(\alpha_i + \theta)$ is the torque induced by $\boldsymbol{u}_i$;

$\boldsymbol{m}_{vi} = \boldsymbol{x}_i \sin(\alpha_i - \theta) - \boldsymbol{y}_i \cos(\alpha_i - \theta)$ is the torque induced by $\boldsymbol{v}_i$;

α_i is the inner normal direction and θ is the half angle of the friction cone.

This system of equations is indeterminate as there are eight unknown scalar variables with only three equations. In search of a solution, contact force at each contact point is set to be either $\boldsymbol{f}_i = \boldsymbol{a}_i \boldsymbol{u}_i$ or $\boldsymbol{f}_i = \boldsymbol{b}_i \boldsymbol{v}_i$, the system can then be successfully solved. Such treatment implies that friction is maximised at all the contacts in the constraint of the workpiece when a feasible solution is found, and accordingly a minimal clamping force can be directly determined.

Therefore, the system of equation (4.3) can be resolved as follows:

$$\begin{pmatrix} \boldsymbol{w}_{1x} & \boldsymbol{w}_{2x} & \boldsymbol{w}_{3x} \\ \boldsymbol{w}_{1y} & \boldsymbol{w}_{2y} & \boldsymbol{w}_{3y} \\ \boldsymbol{m}_1 & \boldsymbol{m}_2 & \boldsymbol{m}_3 \end{pmatrix} \begin{bmatrix} \boldsymbol{c}_1 \\ \boldsymbol{c}_2 \\ \boldsymbol{c}_3 \end{bmatrix} = -\boldsymbol{c}_4 \begin{bmatrix} \boldsymbol{w}_{4x} \\ \boldsymbol{w}_{4y} \\ \boldsymbol{m}_4 \end{bmatrix} - \begin{bmatrix} \boldsymbol{F}_x \\ \boldsymbol{F}_y \\ \boldsymbol{x}_c \boldsymbol{F}_y - \boldsymbol{y}_c \boldsymbol{F}_x \end{bmatrix}$$

or

$$\begin{bmatrix} c_1 \\ c_2 \\ c_3 \end{bmatrix} = \begin{pmatrix} w_{1x} & w_{2x} & w_{3x} \\ w_{1y} & w_{2y} & w_{3y} \\ m_1 & m_2 & m_3 \end{pmatrix}^{-1} \begin{bmatrix} -c_4 w_{4x} - F_x \\ -c_4 w_{4y} - F_y \\ -c_4 m_4 + y_c F_x - x_c F_y \end{bmatrix} \tag{4.4}$$

where bounding wrench $(w_{ix},\ w_{iy},\ m_i)^{\mathrm{T}}$ and its associated scalar c_i is taken from either $(u_{ix},\ u_{iy},\ m_{ui})^{\mathrm{T}}$ and a_i or $(v_{ix},\ v_{iy},\ m_{vi})^{\mathrm{T}}$ and b_i.

Consequently, there are 16 possible equation systems in total corresponding to the permutation of the bounding wrenches. As the solution to Equation (4.4) should be non-negative, a suitable value of c_4 can be derived by setting all the c_1, c_2 and c_3 greater than zero or a positive lower bound. Performing the solution procedure 16 times, the minimal clamping force could be generated as follows:

$$R_4 = \cos(\theta) \min_{i\in[1,2,3,\cdots,16]} \left(c_4(i) \mid c_4(i) \geq 0\right) \tag{4.5}$$

4.2.4 An Illustrating Example

An illustrating example is shown in Figure 4.3 where a workpiece is located with three pins and secured by one clamp. Assuming the friction coefficient is identical for each contact with $\mu = 0.2$, the system of equations (4.3) becomes:

$$\begin{pmatrix} 0.8321 & 0.5547 & -0.1961 & 0.1961 & 0.1961 & -0.1961 \\ -0.5547 & -0.8321 & 0.9806 & 0.9806 & -0.9806 & -0.9806 \\ -1.5254 & -1.2481 & 1.4709 & 1.4709 & -3.3340 & -2.5495 \end{pmatrix} \begin{bmatrix} a_1 \\ b_1 \\ a_2 \\ b_2 \\ a_3 \\ b_3 \end{bmatrix} =$$

$$= \begin{pmatrix} 0.6601 & 0.3204 \\ -0.751 & -0.9473 \\ -2.6346 & -3.0268 \end{pmatrix} \begin{bmatrix} a_4 \\ b_4 \end{bmatrix} + \begin{bmatrix} -150 \\ -50 \\ 125 \end{bmatrix}$$

The minimal clamping force required is found by the developed methodology to be $c_4 \cos\theta = 251.44$N when the lower bound is set to be 10N. Compared with the frictionless case, this clamping force is found to be substantially lower in magnitude. It can be concluded that, as with frictional contacts, higher restraining forces are available for the constraint of the workpiece and hence lower clamping intensity would usually be required.

4.3 Generic Model for Clamping Force Optimisation

The developed methods in the previous sections are only applicable to the planar fixturing case. Concerning the various types of workpieces, a generic model for predicting the minimal clamping magnitude is necessary. Before elaborating on the developed model, assumptions on the workpiece and fixture are indicated as follows:

- The workpiece and the fixturing elements are treated as rigid bodies. The elastic/plastic nature of a workpiece under the influence of clamping and cutting forces is considered in Section 6.3.3.

- Friction exists between the contacting interfaces of the restraints and the workpiece. Coulomb's friction law applies to the contacts. A frictionless assumption is unrealistic as in most fixturing applications, friction is the predominant mechanism for workpiece holding (Lee and Haynes, 1987).

- The contacts between the restraints and the workpiece are represented as unidirectional point contacts. This specifies that only compressive force exists or can be applied in the contacts. Any non-pushing force is physically inadmissible as it suggests the fixturing configuration is no longer valid.

- At each contacting point, at least one of the two engaging surfaces is smooth and therefore the normal direction at that point can be well defined.

The clamping optimisation problem can be stated as: "Given a dimensioned solid model of an arbitrarily-shaped workpiece, configuration of the locators and clamps, cutting tool path and cutting parameters, generate a spectrum of minimum sustainable clamping intensities counterbalancing the influence of varying cutting load while satisfying the requirement that all locators endure strictly positive compressive reaction forces at all times." This is equivalent to total restraint of the workpiece during machining.

4.3.1 Complete Representation of Contact Force

The critical issue in analysing minimum clamping force is the representation of the magnitude and direction of the static friction force at the fixturing contacts. As far as clamping optimisation is concerned, the approximation concept to the friction cone has obvious limitations and is hence no longer satisfactory. Approximation actually diminishes the friction cone; consequently the space of the contact convex is compromised no matter how many sided approximation polyhedrons are used. Although the approximation accuracy can improve with a larger number of polyhedral sides, it would involve excessive computation as the number of sides increases. In order to overcome this shortcoming, a complete representation of the contact force is proposed for the purpose of optimising the clamping force.

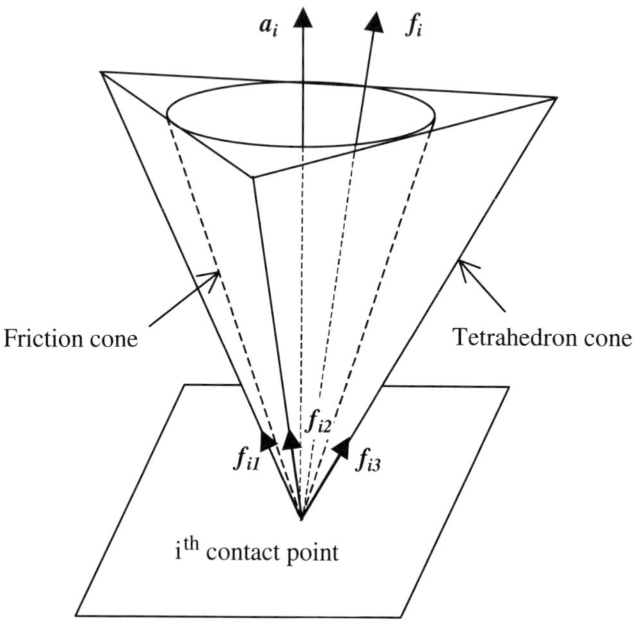

Figure 4.4 Circumscribing tetrahedron cone for contact force representation

As shown in Figure 4.4, a friction cone can be minimally circumscribed by a 3-sided pyramid, a tetrahedron. All its faces are tangent to the friction cone. Assuming the unit generator $\boldsymbol{f}_{ik}$ of the tetrahedron coincides with its edge, then the tetrahedron is easy to define since its generators only need to satisfy the following condition:

$$\boldsymbol{f}_{ik} \cdot \boldsymbol{a}_i = \frac{1}{\sqrt{1+4\mu_i^2}}, \qquad \text{for } k = 1, 2, 3 \tag{4.6}$$

Therefore, the contact force $\boldsymbol{f}_i$ is completely represented by a non-negative linear combination of the generators of the encompassing tetrahedron. At the same time, it must satisfy the following constraint so as to ensure that $\boldsymbol{f}_i$ lies exactly inside the friction cone:

$$f_i = \alpha_{i1} f_{i1} + \alpha_{i2} f_{i2} + \alpha_{i3} f_{i3} \tag{4.7}$$

$$\left\| \alpha_{i1} f_{i1} + \alpha_{i2} f_{i2} + \alpha_{i3} f_{i3} \right\| \leq \sqrt{\frac{1+\mu_i^2}{1+4\mu_i^2}} \left(\alpha_{i1} + \alpha_{i2} + \alpha_{i3} \right) \tag{4.8}$$

Consequently, the contact force f_i can be resolved to an equivalent wrench w_i consisting of a force and a moment about the origin of the reference frame R:

$$w_i = \begin{bmatrix} f_i \\ r_i \times f_i \end{bmatrix} = \begin{bmatrix} \alpha_{i1} f_{i1} + \alpha_{i2} f_{i2} + \alpha_{i3} f_{i3} \\ \alpha_{i1}(r_i \times f_{i1}) + \alpha_{i2}(r_i \times f_{i2}) + \alpha_{i3}(r_i \times f_{i3}) \end{bmatrix} \tag{4.9}$$

where $\alpha_{ik} \geq 0$.

The merit of the new representation of contact force is obvious. It is an exact representation approach, no approximation is involved here so that the contact convex is kept intact without any loss of space. Moreover, fewer variables α_{ik} are involved as only three generators are needed for each contact force, resulting in reduced computation complexity for optimisation.

4.3.2 Positive Location Constraint

It is noted that clamps can only exert bounded forces depending on their mechanism, friction, deformability and so on. The type of contacts between the restraints and the workpiece is also restrictive to the force that can exist in these interfaces. Therefore, the clamping force optimisation scheme must subject to a set of constraints for all relevant physical phenomena and one of them is the positive locating requirement.

Since the contact is unidirectional and the restraints can only push (not pull) the workpiece, the normal vector at the contact point is initially defined as directed into the workpiece. Therefore, the normal force for each contact point must remain strictly positive to ensure that the workpiece stays in contact with the restraints because the negative values are physically impossible.

In a practical workpiece-fixture system, the reaction forces on the locators must be held within an appropriate positive range. If the reaction force is greater than an upper bound, surface yielding may occur to the locator due to high contacting stress and its locating accuracy may be lost. On the other hand, if the reaction force drops to zero, that means the workpiece has been detached from the locator and its reference datum may be gone. Consequently, it is reasonable to constrain the reaction forces to be between lower and upper bound values. Since the objective is to minimise the clamping intensities, upper bound constraint on the reaction forces is redundant. Therefore, the positive reaction force constraints can be expressed as follows:

$$\boldsymbol{f}_i \cdot \boldsymbol{a}_i > 0$$

or

$$\sum_{k=1}^{3} \frac{\alpha_{ik}}{\sqrt{1+4\mu_i^2}} > 0 \tag{4.10}$$

Minimum clamping force determination is a constrained optimisation procedure. Any optimisation problem subject to *open* constraints such as Equation (4.10) usually renders no solutions in numerical computation. Examine a simple example of `Minimise f(x)=2-x subject to 0<x<1`. It is known that the problem has a lower bound of one. It is difficult to achieve the minimum in numerical computation due to the open constraint. Consequently, a non-zero lower bound LB_i, say 10 N, is specified for the reaction forces on the locators so that the open inequality is replaced by a closed form:

$$\sum_{k=1}^{3} \frac{\alpha_{ik}}{\sqrt{1+4\mu_i^2}} \geq \boldsymbol{LB}_i \qquad \text{for all locators} \tag{4.11}$$

4.3.3 External Load

Given the cutter geometry, machining parameters and workpiece material property, the machining forces can be estimated using the established force prediction model (Kline, 1982). When an end milling operation is performed on a workpiece, the instantaneous resultant milling force $\boldsymbol{F}(t)$ and a couple $\boldsymbol{M}(t)$ are imposed on the workpiece at time t (Figure 4.5). $\boldsymbol{F}(t)$ and $\boldsymbol{M}(t)$ are defined with respect to a cutting tool reference frame. Let $\boldsymbol{r}(t)$ be the position of the origin of the tool reference frame about the fixed reference frame R. Consequently the machining forces can be represented by a cutting wrench about the fixed reference frame as follows:

$$\boldsymbol{w}(t) = \begin{bmatrix} \boldsymbol{F}(t) \\ \boldsymbol{M}(t) + \boldsymbol{r}(t) \times \boldsymbol{F}(t) \end{bmatrix} \tag{4.12}$$

Note that the milling forces are cyclic in nature and $\boldsymbol{r}(t)$ is always changing as the cutter moves along its path. For conciseness, only a finite number of points along the tool path are considered in the computation and every point is assumed to experience the peak cutting forces.

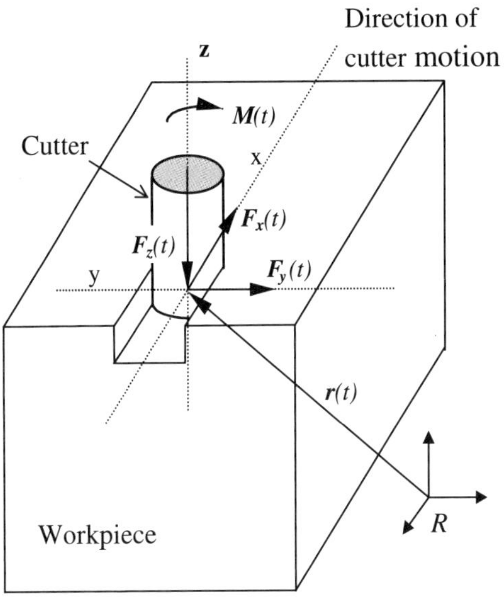

Figure 4.5 Forces and torque acting on the workpiece during end milling

4.3.4 Workpiece Static Equilibrium

During the machining process, the workpiece is subjected to a set of holding wrenches exerted by the fixturing contacts, interaction with the cutting tool and gravitational forces. It is required that all the wrenches acting on the workpiece must sum to zero if the workpiece is to be in a state of static equilibrium. The following static equilibrium condition must be satisfied throughout the operation:

$$\sum_{i=1}^{n} \boldsymbol{w}_i + \boldsymbol{w}_g + \boldsymbol{w}(t) = 0 \tag{4.13}$$

where $\boldsymbol{w}_i$ is the holding wrenches generated by the restraints, and $\boldsymbol{w}_g$ the wrench induced by the gravity and $\boldsymbol{w}(t)$ the external load caused by the processing forces.

Substituting Equation (4.9) and Equation (4.12) into Equation (4.13), the following expression can be obtained:

$$\begin{bmatrix} \sum_{i=1}^{n}\sum_{k=1}^{3}\alpha_{ik} f_{ik} \\ \sum_{i=1}^{n}\sum_{k=1}^{3}\alpha_{ik}\left(r_i \times f_{ik}\right) \end{bmatrix} + \begin{bmatrix} F(t) \\ M(t) + r(t) \times F(t) \end{bmatrix} + w_g = 0 \tag{4.14}$$

Let the centre of gravity of the workpiece be $r_c = [x_c \quad y_c \quad z_c]^T$, the gravitational force be $F_c = [0 \quad 0 \quad -mg]^T$, then the wrench associated with the self-weight can be found as follows:

$$w_g = [0 \quad 0 \quad -mg \quad -mgy_c \quad mgx_c \quad 0]^T$$

4.4 Optimal Clamping Force Algorithm

In several previous fixturing configuration verification systems (Asada and By, 1985, Chou *et al.*, 1989), the assumption of frictionless contacts between the workpiece body and fixturing elements is made. This results in simpler mathematical modelling and consequently less effort in generating a feasible solution.

However, the frictionless approach is a poor approximation to the workholding problem in practice. It not only severely confines the solution space, but also simply cannot characterise some of the fixturing problems. One obvious example is the case of a 2-D workpiece held at antipodal points by two fixturing elements, which is impossible without friction. Another example is that a flat component on a surface grinding machine table is solely constrained by friction generated by a powerful electro-magnetic chuck. In conclusion, a realistic optimisation model must include the effect of friction. The existence of friction in a model makes the solution space broader, and at the same time more interesting and intricate.

4.4.1 Minimum Clamping

The objective is to determine the minimum clamping forces that guarantee the workpiece would not slip and/or detach during the machining process. The associated difficulty is that the problem of workpiece static equilibrium governed by Equation (4.14) is statically indeterminate. Consequently, a numerical constrained-search algorithm should be employed to compute the optimal clamping forces based on an appropriate criterion for a given set of restraints.

The present criterion is formulated based on an intuitive idea that *smaller* compressive forces are required in order to grip an object in the presence of *larger* friction. Imagine three fingers are used to prehend a smooth object (lower friction) and a rough surface object (higher friction) respectively, it is heuristically clear that less articulated forces are needed to grasp the latter firmly. Accordingly, friction is considered to have a profound *beneficial* effect on the problem of workpiece constraint within a fixture. This seems to coincide with the fact that many conventional fixtures, such as vises and chucks, mainly depend on friction to maintain the static equilibrium and stability of workpieces. It is a natural extension that if the effect of friction can be maximised on the workpiece constraint, the required clamping force will be a minimum.

Proposition 4.1 When friction acting on a workpiece in an n-restraint fixturing configuration reaches the maximum, the magnitude of the clamping intensity required to achieve static equilibrium is the minimum.

This proposition may be referred to as the maximum friction capacity criterion. Consider at the point of contact C_i between the workpiece and the restraint, the contact force $\boldsymbol{f}_i$ can be decomposed into two components: normal force $\boldsymbol{f}_i^N$ and tangential force $\boldsymbol{f}_i^T$. The magnitude of the tangential component can be represented as $\left\|\boldsymbol{f}_i^N\right\| = \omega_i \mu_i \left\|\boldsymbol{f}_i^N\right\|$, the friction at C_i is hence maximised as $\omega_i \Rightarrow 1$. Consequently, ω_i may be defined as the friction capacity ratio:

$$\omega_i = \frac{\left\|\boldsymbol{f}_i \times \boldsymbol{a}_i\right\|}{\mu_i (\boldsymbol{f}_i \cdot \boldsymbol{a}_i)}$$

Substituting Equation (4.7) into above expression yields:

$$\omega_i = \sqrt{4 + \mu_i^{-2}} \frac{\left\| \sum_{k=1}^{3} \alpha_{ik} \left(\boldsymbol{f}_{ik} \times \boldsymbol{a}_i \right) \right\|}{\sum_{k=1}^{3} \alpha_{ik}} \tag{4.15}$$

Now, the objective is shifted to maximise the summation of the friction capacity ratio ω_i in order to generate a set of minimum necessary actuating intensities. This task can be achieved by solving the following constrained non-linear programming problem:

$$\text{Maximise} \qquad \sum_{i=1}^{n} \sqrt{4 + \mu_i^{-2}} \frac{\left\| \sum_{k=1}^{3} \alpha_{ik} \left(\boldsymbol{f}_{ik} \times \boldsymbol{a}_i \right) \right\|}{\sum_{k=1}^{3} \alpha_{ik}}$$

Subject to

(a) $$\begin{bmatrix} \sum_{i=1}^{n}\sum_{k=1}^{3}\alpha_{ik} f_{ik} \\ \sum_{i=1}^{n}\sum_{k=1}^{3}\alpha_{ik}(r_i \times f_{ik}) \end{bmatrix} + \begin{bmatrix} F(t) \\ M(t)+r(t)\times F(t) \end{bmatrix} + w_g = 0$$

(b) $$\sqrt{\frac{1+\mu_i^2}{1+4\mu_i^2}}(\alpha_{i1}+\alpha_{i2}+\alpha_{i3}) - \|\alpha_{i1} f_{i1} + \alpha_{i2} f_{i2} + \alpha_{i3} f_{i3}\| \geq 0$$

for i = 1, 2, ..., n

(c) $$\sum_{k=1}^{3}\frac{\alpha_{ik}}{\sqrt{1+4\mu_i^2}} - LB_i \geq 0$$

for i = 1,2, ..., m

(d) $\alpha_{ik} \geq 0$

for i = 1, 2, ..., n; k = 1, 2, 3

4.4.2 Safety Consideration

In a manufacturing environment, the physical phenomenon of friction between the workpiece and the fixturing elements is affected by many factors such as surface cleanliness, presence of cutting fluid and foreign particle inclusions, etc. In other words, the coefficient of friction μ is physically not well defined and is of stochastic nature to some extent. A clamping intensity scheme generated by the proposed algorithm may cause the workpiece to be in a critical stability where one or more contact points are at the verge of slippage with $\omega_i \Rightarrow 1$. This state of critical stability may be disrupted by any external disturbances. Therefore, a safety factor s is introduced as follows:

$$\mu = s\mu_0 \text{, with } 0 < s \leq 1 \tag{4.16}$$

where μ_0 is the experimental estimate of the coefficient of static friction from inclination tests.

4.4.3 Maximum Clamping

In practice, it may be required that the clamping intensity remains within a particular range. Restriction of clamping force to a minimum value helps to reduce the chance of insufficient clamping. On the other hand, a maximum limit helps to reduce the risk of over-clamping, and hence the unfavourable and large deformation of the workpiece can be avoided by restricting excessive clamping forces. In short, minimum and maximum clamping forces would provide a useful guideline for machinists when it comes to planning proper clamping operations.

The maximum clamping force may be defined as the clamping intensity that can achieve static equilibrium when friction at the fixturing contacts is least relied upon for counteracting the external load. Therefore, the proposed minimum clamping algorithm can be readily modified for the computation of maximum clamping intensity. Instead of a maximisation procedure, it becomes a minimisation problem as follows:

Minimise $\sum_{i=1}^{n} \omega_i$

It is subject to the same constraints except the constraint (c), which is replaced by a set of inequality constraints on all clamps as follows:

$$\boldsymbol{UB}_i - \sum_{k=1}^{3} \frac{\alpha_{ik}}{\sqrt{1+4\mu_i^2}} \geq 0 \tag{4.17}$$

where the UB_i is the largest possible clamping force a clamp could provide or sustain.

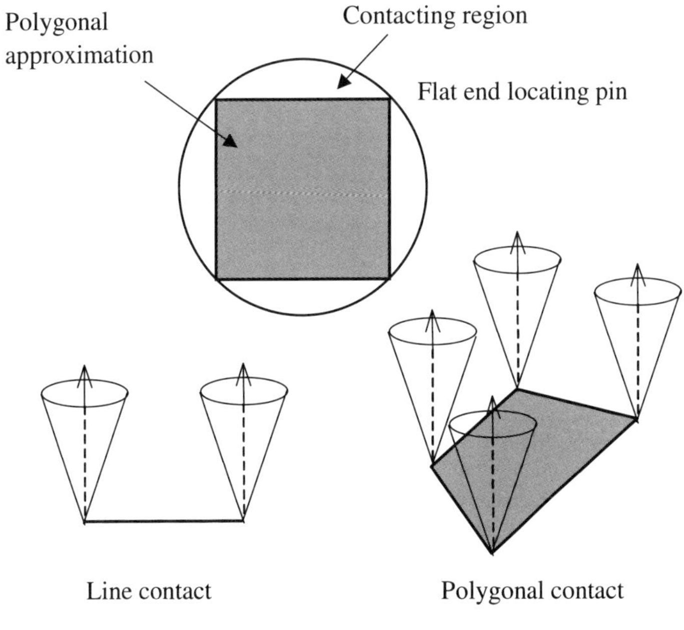

Figure 4.6 Contacts and approximations

4.4.4 Polygonal Contacts

The proposed algorithm is readily extendable to polygonal contacts, although point contacts with friction are assumed during the model development. Other contact types such as line and surface contacts are commonly involved in fixtures due to various sizes and shapes of workpieces and fixturing elements. In general, the contacts between a workpiece and a restraint may be represented by contacts between any two topological elements of the objects as follows:

$$\boldsymbol{contact} = \{(\boldsymbol{e}_1, \boldsymbol{e}_1')(\boldsymbol{e}_2, \boldsymbol{e}_2')\cdots(\boldsymbol{e}_n, \boldsymbol{e}_n')\} \tag{4.18}$$

where a pair $(\boldsymbol{e}_i, \boldsymbol{e}_i')$ represents an elemental contact between topological elements $\boldsymbol{e}_i$ and $\boldsymbol{e}_i'$ of the workpiece and restraint respectively. The topological elements are the components of a boundary-represented solid model such as vertices, straight and curved edges, planar and curved faces.

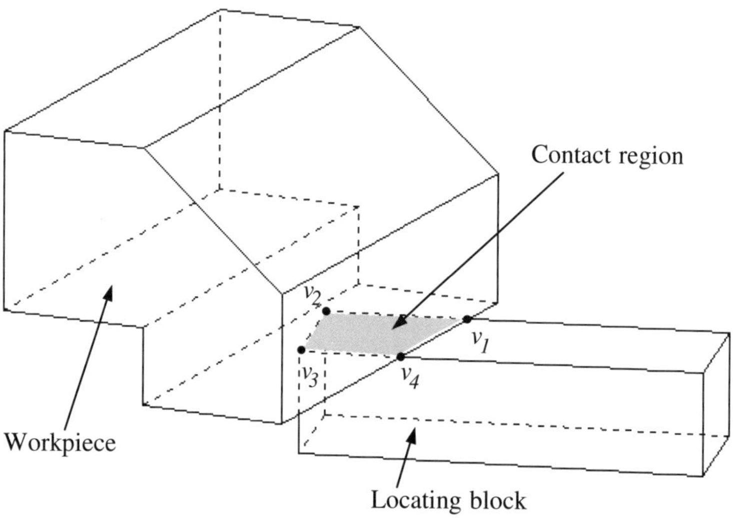

Figure 4.7 Contact forces at the marked vertices of the shaded contact region

Fixturing contacts can be approximated by either line contacts or polygonal contacts (Figure 4.6). The contacting wrench associated with a convex polygonal contact can be represented as a non-negative combination of the wrenches exerted at the extreme points of the polygon. Therefore, line contacts can be represented by a combination of two point contacts at the end points of the line segments. Surface contacts can be approximated by a combination of point contacts at the vertices of a polygon representing the planar contacts as shown in Figure 4.6.

As depicted in Figure 4.7 where a polygonal contact region occurs between a workpiece and a locating block, only the contact forces at the vertices of the contact region need to be modelled. Supposing the contacting polygon has v vertices, the contacting wrench, $\boldsymbol{w}$, associated with the contact area can be expressed as follows:

$$\begin{aligned}\boldsymbol{w} &= \sum_{i=1}^{v} \xi_i \begin{bmatrix} \boldsymbol{f}_i \\ \boldsymbol{r}_i \times \boldsymbol{f}_i \end{bmatrix} \\ &= \sum_{i=1}^{v} \begin{bmatrix} \alpha'_{i1}\boldsymbol{f}_{i1} + \alpha'_{i2}\boldsymbol{f}_{i2} + \alpha'_{i3}\boldsymbol{f}_{i3} \\ \alpha'_{i1}(\boldsymbol{r}_{i1} \times \boldsymbol{f}_{i1}) + \alpha'_{i2}(\boldsymbol{r}_{i2} \times \boldsymbol{f}_{i2}) + \alpha'_{i3}(\boldsymbol{r}_{i3} \times \boldsymbol{f}_{i3}) \end{bmatrix}\end{aligned} \tag{4.19}$$

Therefore, the presented methodology is readily applicable to the fixturing problems with polygonal support faces.

4.5 Algorithm Implementation

The design variables are the multiplier scalars α_{ik} for the generating vectors associated with each contact. For a workpiece undergoing n constraints by restraints, if each friction cone at the contacting point is encompassed by a tetrahedron, then there are $3n$ unknown scalar variables in the problem. The task is to search for a feasible solution in the solution space delimited by the proposed constraints, which achieves a maximum objective function value. Both the objective and constraints are non-linear functions of the design variables.

General non-linear programming methods are available for solving such a problem. One of the approaches is Sequential Quadratic Programming (SQP) which can perform efficiently and provide accurate solutions (Powell, 1978; Schittowski, 1985). In the SQP method, a Quadratic Programming (QP) sub-problem is solved at each iteration. An estimate of the Hessian of the Lagrangian is updated at each iteration using a quasi-Newton updating method. In this study, the optimisation algorithm is implemented on the Unix platform. The operational procedure of this methodology is illustrated in Figure 4.8.

The optimisation procedure starts at the origin and yields a set of optimal variables α_{ik}, from which both reaction (clamping) and friction force values can be derived for locators (clamps). In general, the optimisation program runs smoothly and always converges to a solution with optimum values returned in a fraction of a second in 2-D cases or within a minute in 3-D cases depending on the fixture configuration complexity. The existence of a feasible solution is secured because the fixturing scheme is verified in such a way as to guarantee force closure for the workpiece by the verification procedure in Chapter 3. Results of two case studies are discussed in the following sections.

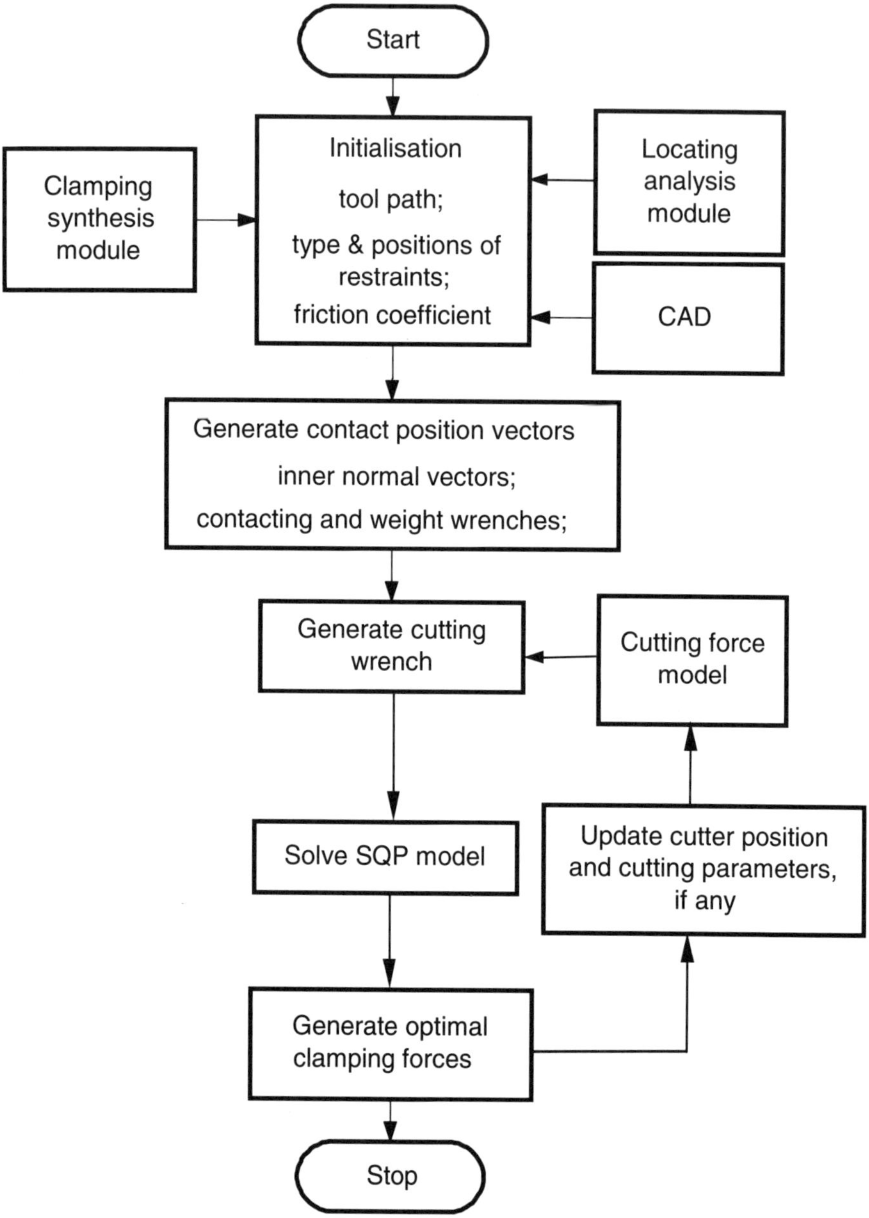

Figure 4.8 Flowchart of dynamic clamping force generation

4.6 Case Studies

4.6.1 Planar Fixturing

A 2-D case study is first explored where a general 2-D workpiece is secured by three locators and one clamp as shown in Figure 4.9. The locators are spherical buttons and the clamp also has a spherical tip, and hence making point contacts with the object on its sides. The coefficient of static friction is reasonably assumed to be 0.2 for interfaces between the steel fixturing elements and the aluminium workpiece. An external load of $[\boldsymbol{F}_x, \boldsymbol{F}_y, \boldsymbol{T}_z]^{\mathrm{T}}$ = [250N, -300N, 3000N·mm]^{T} is imposed on the workpiece, which transverses from starting point a to end point b during machining. It is required to generate the optimum clamping force needed to balance this dynamic machining load.

First of all, the fixturing configuration has to be verified. An initial fixturing layout with the clamp positioned at point 1 is shown in Figure 4.9(a). After checking for both force closure and clamping equilibrium using the algorithm in Chapter 3, it is found that the clamp is not placed within the valid clamping region (in bold line) and it is hence not a viable configuration and the fixturing scheme has to be redesigned. Finally, the clamp is chosen to be at point 2 as illustrated in Figure 4.9(b) according to the ranking metric of the radius of the maximum inscribed sphere within the convex hull of the bounding wrenches. Now this fixturing configuration is verified to be capable of achieving both force closure and clamping equilibrium, and thus guaranteeing the workpiece stability.

The cutting wrench is calculated at 41 equidistant points along the tool path, which translates into 41 dynamic fixturing solutions. The minimum clamping force required over time is shown in Figure 4.10, together with the reaction forces on three locators. The clamping intensities are just enough to keep the workpiece in a state of static equilibrium while providing all positive reaction forces for the locators all the time. The highest clamping force needed is 654 N at the start of the process while the lowest clamping force is 79N. If a dynamic clamp is employed, then the clamp should provide variable clamping intensities with respect to the positions of the cutter. If a static clamp is used, the highest clamping force of 654N must be applied and maintained throughout the machining operation.

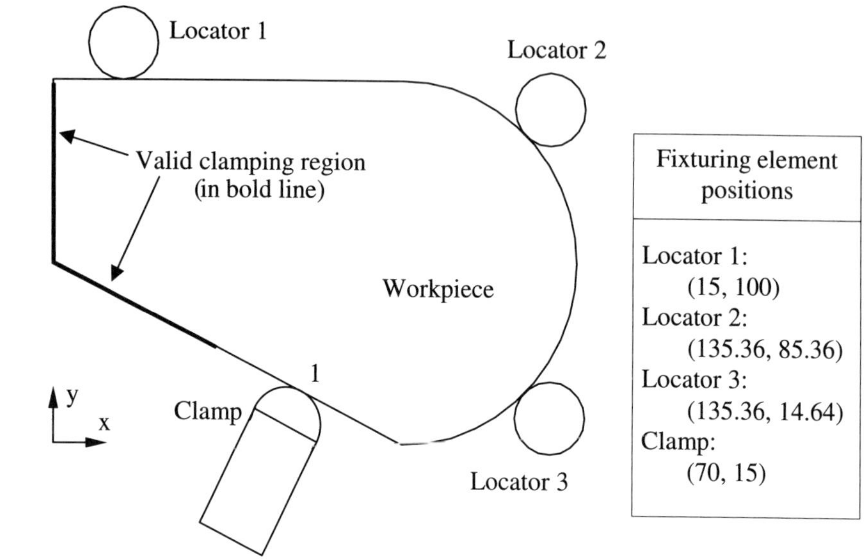

(a) Invalid fixturing configuration

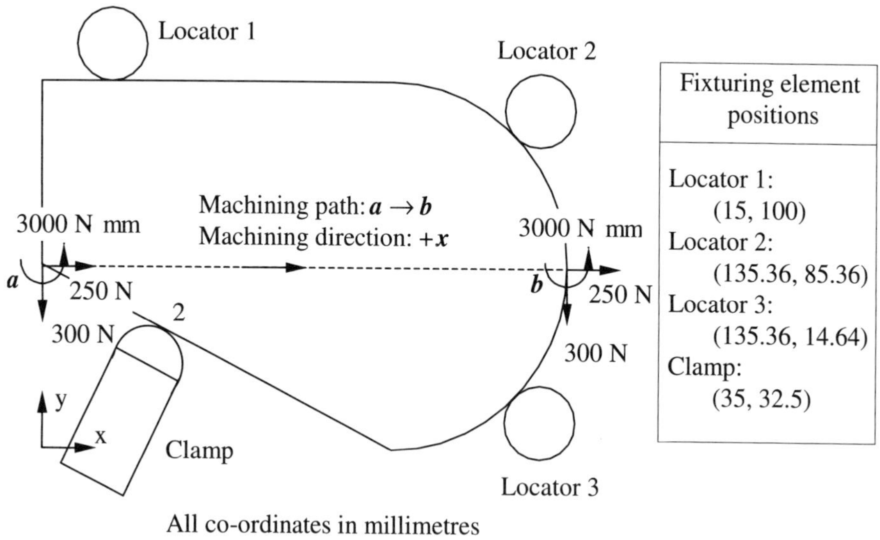

(b) Valid fixturing layout

Figure 4.9 Case study of fixturing configuration for a 2-D workpiece

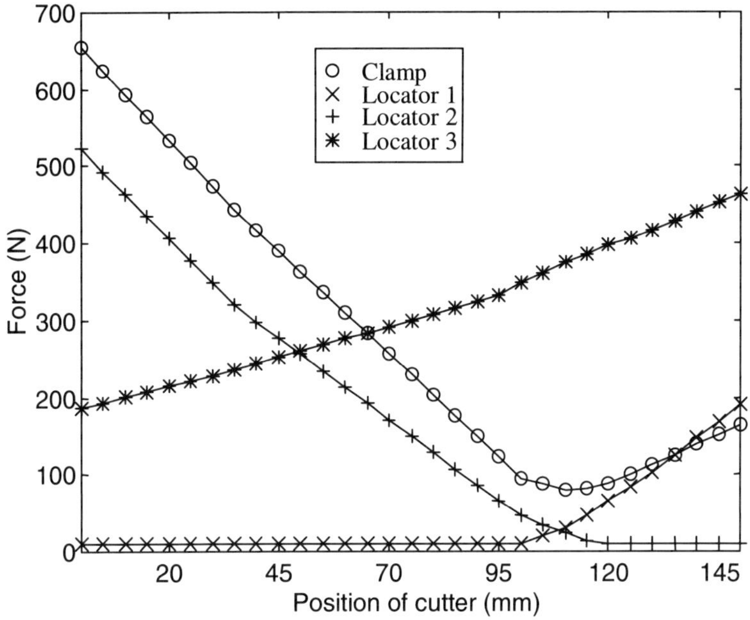

Figure 4.10 Generated clamping and reaction forces

4.6.2 Spatial Fixturing

The second case study is a more complex 3-D fixturing problem. As shown in Figure 4.11, the workpiece is a prismatic block with a step feature on its top face. A slot milling operation is to be performed on the workpiece so as to produce a through slot. The workpiece is located by six spherically tipped locating pins according to the 3-2-1 locating principle and secured by two side clamps. The clamping points *C1* and *C2* are positioned at (110, 0, 60) and (0, 60, 60) respectively. The locations and orientations of the locators are tabulated in Table 4.1. There are 24 bounding wrenches in six-dimension associated with the fixturing contacts. The fixturing configuration is verified to be viable in terms of force closure and clamping equilibrium.

The initial depth of cut is 1 mm and the subsequent depth is increased to 2 mm. Milling forces can be estimated using proven empirical cutting force formula with results tabulated in Table 4.2. The body force for the aluminium workpiece is $[0, 0, -90\text{N}]^T$ at (110, 60, 47.5), its centroid. The coefficient of static friction is assumed to be identical for all the contacting points with $\mu = 0.2$.

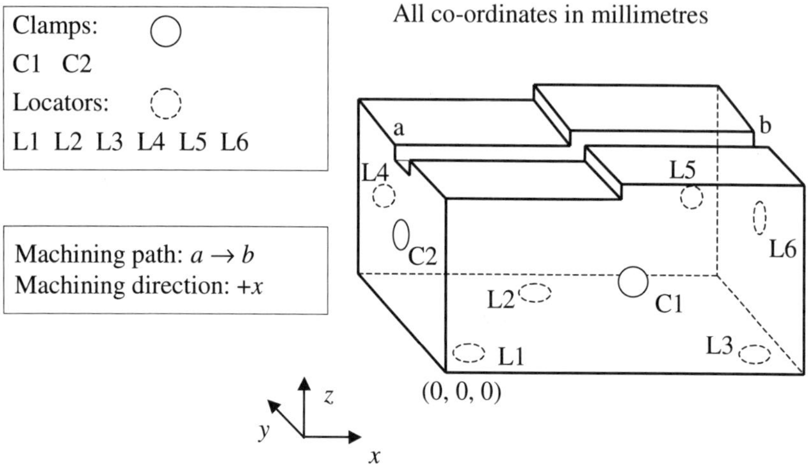

Figure 4.11 A case study of 3-D fixturing configuration

Table 4.1 Positions and orientations of the fixturing contacts

	Positions of the contact points	Inner normal at the contacts
Clamps	C1 (110, 0, 60)	C1 $[0, 1, 0]^T$
	C2 (0, 60, 60)	C2 $[1, 0, 0]^T$
Locators	L1 (10, 10, 0)	L1 $[0, 0, 1]^T$
	L2 (110, 110, 0)	L2 $[0, 0, 1]^T$
	L3 (210, 10, 0)	L3 $[0, 0, 1]^T$
	L4 (10, 120, 60)	L4 $[0, -1, 0]^T$
	L5 (210, 120, 60)	L5 $[0, -1, 0]^T$
	L6 (220, 60, 60)	L6 $[-1, 0, 0]^T$

Table 4.2 Estimated milling forces

Components	Cutter's position $0 \le x < 110$ mm	Cutter's position $110 \text{ mm} \le x \le 220$ mm
x-direction	55 N	110 N
y-direction	-100 N	-200 N
z-direction	-25 N	-50 N
Torque	-1250 N·mm	-2500 N·mm

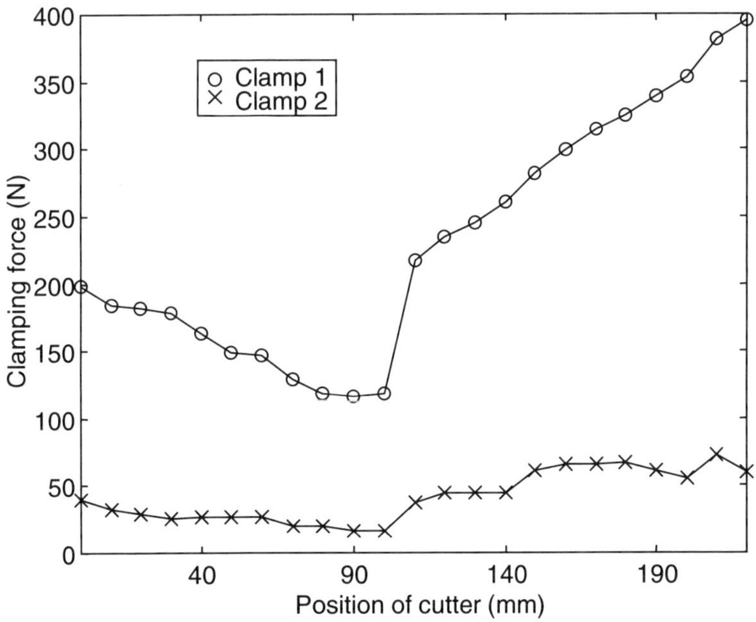

Figure 4.12 Clamping forces needed to constrain the workpiece over time

The clamping forces required to counterbalance the dynamic milling forces are calculated using the optimisation algorithm and illustrated in Figure 4.12. The required clamping intensity for clamp *C1* steadily decreases from 198N to 116N as the cutter moves on during the first half of the path. Less clamping forces are needed when the cutter is travelling closer to the middle of the path. When the external load steps up as a bigger depth of cut is engaged, the clamping intensity correspondingly goes up from 116N to 217N. During the second half of the path, as the cutter transverses away from the clamp, the required clamping intensity gradually increases to 395N.

The trend of the clamping intensity for clamp *C2* is almost the same as that for *C1*, but substantially smaller in magnitude. Less clamping force is required due to the fact that the *x*-directional component of the cutting forces is directed towards a fixed restraint, locator *L6*. The clamping intensity goes down slightly from 39N to 16N during the first half of the path, while on the other hand it increases from 37N to 72N for the second half of the path.

The reaction forces on the locators under variable clamping forces are shown in Figure 4.13. All the locators undergo a strictly positive reaction force all the time, implying that workpiece stability is always maintained during the process by the variable clamping intensities. The maximum reaction force of 249N is experienced by locator *L6* as the cutter moves to its end point. Reaction forces do not drop below 10N as a lower bound of same value is imposed in the optimisation scheme. A constant, rather than a variable, clamping force setting can always be employed just for easiness. It is clear that the workpiece stability must be maintained when the static clamping intensities for clamps *C1* and *C2* are set to be 395N and 72N respectively.

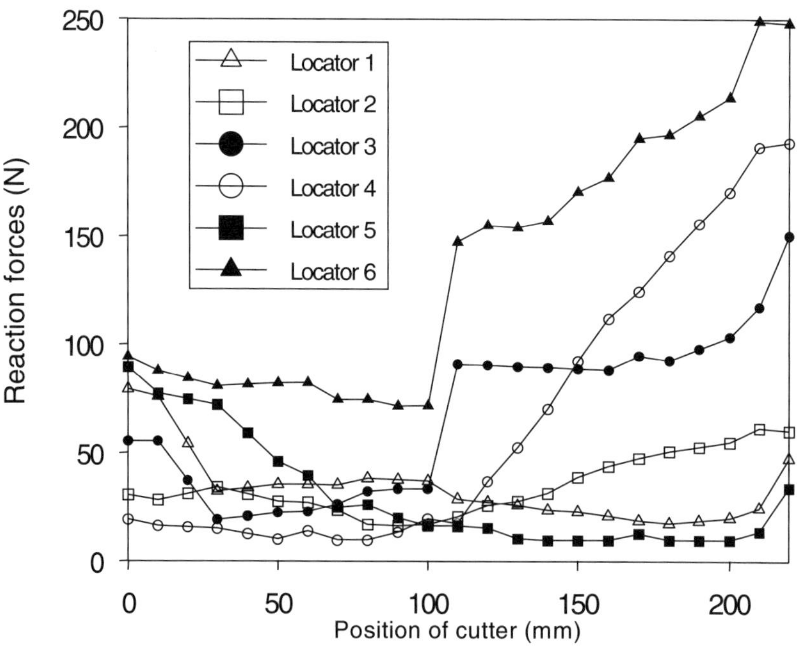

Figure 4.13 Reaction forces on locators for milling pass over time

For comparison, the clamping forces required to keep the workpiece in stability when friction is least dependent on are illustrated in Figure 4.14. They are higher in magnitude than those needed under maximum friction dependence. The highest clamping forces for clamps *C1* and *C2* are 492N and 232N respectively. The maximum and minimum clamping intensities serve as guides in practical workholding. Under maximum clamping conditions, the workpiece is least likely to slip or detach within a fixture during machining. Another conclusion may be drawn from the results is that clamping force should be increased when it comes to fixturing a smooth workpiece, otherwise the processing force should be kept as low as possible.

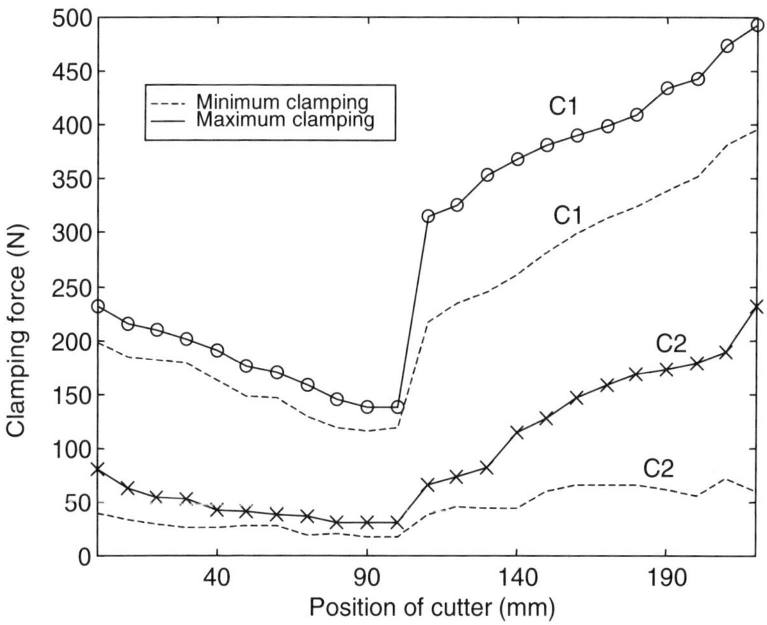

Figure 4.14 Maximum vs. minimum clamping forces

4.7 Summary

In this chapter, both direct computation and non-linear programming approaches are presented for the generation of minimal sustainable clamping forces. The new direct computation method provides a straightforward solution for planar fixture planning by which the valid clamping intensity can be arrived at quickly.

The non-linear programming procedure is useful for the optimisation of fixturing configurations in general. Polyhedral cone approximation is no longer used in minimum clamping intensity determination. A complete representation of contact force is presented, which provides a mechanism for robust and complete solution. A maximum friction capacity criterion is proposed in order to determine the minimum clamping intensities. The frictional methodology allows a sound solution to clamping intensity setting since there are more restraining forces available.

4.8 Bibliography

Asada, H. and A.B. By. Kinematic Analysis of Workpart Fixturing for Flexible Assembly with Automatically Reconfigurable Fixtures. *IEEE Journal of Robotics and Automation*, 1(2), pp. 86-94. 1985.

Chou, Y.-C., V. Chandru and M.M. Barash. A Mathematical Approach to Automatic Configuration of Machining Fixtures: Analysis and Synthesis. *Journal of Engineering for Industry*, 111(4), pp.579-584. 1989.

Cogun, C. The Importance of the Application Sequence of Clamping Forces on Workpiece Accuracy. *Journal of Engineering for Industry*, 114, pp.539-543. 1992.

DeMeter, E.C. Restraint Analysis of Fixtures which Rely on Surface Contact. *Journal of Engineering for Industry*, 116(2), pp. 207-215. 1994.

DeMeter, E.C., W. Xie, S. Choudhuri, S. Vallapuzha and M.W. Trethewey. A Model to Predict Minimum Required Clamp Pre-loads in Light of Fixture-Workpiece Compliance. *International Journal of Machine Tools and Manufacture*, Vol.41, No.7, pp.1031-1054. 2001.

Fuh, J.Y.H. and A.Y.C. Nee. Verification and Optimisation of Workholding Schemes for Fixture Design. *Journal of Design and Manufacturing*, 4, pp.307-318. 1994.

Jeng S.L., L.G. Chen and W.H. Chieng. Analysis of Minimum Clamping Force. *Int. J. Mach. Tools Manufact.*, 35(9), pp.1213-1224.1995.

Kerr, J. and B. Roth. Analysis of Multi-fingered Hands. *Int. J. Robotics Research*, 4(4), pp.3-17. 1986.

Kline, W.A. *The Prediction of Cutting Forces and Surface Accuracy for the End Milling Process*. PhD Thesis, University of Illinois at Urbana-Champaign. 1982.

Krishnakumar, K. and S.N. Melkote. Machining Fixture Layout Optimization Using the Genetic Algorithm. *International Journal of Machine Tools and Manufacture*, Vol.40, No.4, pp.579-598. 2000.

Lakshminarayana, K. *Mechanics of Form Closure*. Technical report 78-DET-32, ASME. 1978.

Lee, J.D. and L.S. Haynes. Finite-Element Analysis of Flexible Fixturing System. *Journal of Engineering for Industry*, 109(2), pp.134-139. 1987.

Lee, S.H. and M.R. Cutkosky. Fixturing Planning with Friction. *Journal of Engineering for Industry*, 113, pp. 320-327. 1991.

Li, B. and S.N. Melkote. Optimal Fixture Design Accounting for the Effect of Workpiece Dynamics. *Int. J. Adv. Mfg. Technol.*, Vol. 18, pp.701-707. 2001.

Mishra, B. and N. Silver. Some Discussion of Static Gripping and its Stability. *IEEE Sys. Man Cybernet.*, 19(4), pp. 783-796. 1989.

Mittal, R.O., P.H. Cohen and B.J. Gilmore. Dynamic Modelling of the Fixture-Workpiece System. *Robotics and Computer-Integrated Manufacturing*, 8(4), pp.201-217. 1991.

Nguyen, V.D. Constructing Force-closure Grasps. *Int. J. of Robotics Research*, 7(3), pp.3-16. 1988.

Nnaji, B.O., S. Alladin and P. Lyu. A Framework for a Rule-base Expert Fixturing System for Face Milling Planar Surface on a CAD System Using Flexible Fixtures. *Journal of Manufacturing System*, 7(3), pp.193-207. 1988.

Powell, M.J.D. A Fast Algorithm for Nonlinearly Constrained Optimisation Calculations. In *Numerical Analysis*, Vol.630, ed. by G.A. Watson. Springer Verlag. 1978.

Salisbury, J.K. and B. Roth. Kinematics and Force Analysis of Articulated Hands. *J. Mechanisms Transmissions Automation Design*, 105, pp.33-41. 1982.

Schittowski, K. NLQPL: A FORTRAN-subroutine Solving Constrained Non-linear Programming Problems. *Operations Research*, 5, pp.485-500. 1985.

Trappey, A.J.C. and C.R. Liu. An Automatic Workholding Verification System. *Robotics and Computer-integrated Manufacturing*, 9(4/5), pp.321-326. 1992.

Vallapuzha, S., E.C. DeMeter, S. Choudhuri and R.P. Khetan. An Investigation into the Use of Spatial Coordinates for the Genetic Algorithm-based Solution of the Fixture Layout Optimization Problem. *Int. J. Mach. Tools. Mfg.*, Vol.42, pp. 265-275. 2002.

Vallapuzha, S., E.C. DeMeter, S. Choudhuri and R.P. Khetan. An Investigation of the Effectiveness of the Fixture Layout Optimization Method. *Int. J. Mach. Tools Mfg.*, Vol.42, pp251-263. 2002.

Yeh, J.H. and F. W. Liou. Contact Condition Modelling for Machining Fixture Setup Processes. *International Journal of Machine Tools and Manufacture*, Vol.39, No.5, pp.787-803. 1999.

Chapter 5

Workpiece-Fixture Interaction

5.1 Introduction

The dynamic characteristics of a workpiece-fixture system play an important role in determining the machining accuracy of a workpiece. The contact interaction between a workpiece and the fixturing elements and the deflection of workpiece induced by the contact forces as well as external loads need to be explored. As reaction forces between the contact interfaces are the direct consequence of the cutting and the clamping forces imparted to the workpiece in addition to the inherent effect of the weight of the workpiece, they are reflective of the temporal interactive relationship between the workpiece and the fixturing elements and provide a good understanding of the dynamic force modelling of the workpiece-fixture system. In this chapter, a developed sensor-integrated fixture for experimentally investigating the force interaction is discussed. The finite element analysis is next employed to investigate the interaction of workpiece/fixture combination, to determine the deflection of a workpiece under given loads and the effects of dynamic clamping on the deflection. The developed system forms a sound foundation for an intelligent fixture as it can be used to execute on-line monitoring and evaluation of the fixturing performance.

5.2 Experimental Investigation

This section presents a methodology for measuring and analysing the interaction between a workpiece and its fixturing elements. A sensor-integrated fixture system, in which real-time reaction forces on locators can be measured through a data acquisition system, has been developed for end milling operations. Reaction forces with respect to cutting forces, location of cutter and clamping intensities are monitored throughout the machining process and a spectral analyser is used to investigate the dynamic behaviour of the workpiece-fixture interaction.

5.2.1 Sensor-integrated Fixture System

5.2.1.1 Forces within a Fixture

A good understanding of all the ambient forces within a fixture is a prerequisite necessary to identify significant forces that can be used to indicate the workpiece stability. Six different types of forces acting within the workpiece-fixture can be summarised as follows:

- *Gravitational force*. The inherent gravitational weight can usually be ignored as its contribution to the deflection of the workpiece is much smaller than the cutting and clamping forces. For a structurally delicate workpiece, however, it should be taken into consideration.

- *Cutting force*. The cutting forces are generated during the machining process and should be carefully counterbalanced by sufficient clamping. The force values may be measured in real-time using a dynamometer.

- *Clamping forces*. Indispensable to a viable workholding scheme, they are applied by the active elements of a fixture. Applied clamping forces can be estimated using a torque wrench as in the case of screw clamps, or using the pressure gauge readings when hydraulically or pneumatically operated clamps are in use.

- *Reaction forces on locators*. They are not independent parameters and are determined by the external forces acting on the workpiece.

- *Frictional force*. In most cases, frictional force exists between the fixturing elements and the workpiece and plays a significant role in resisting a workpiece from slippage. In a valid fixture configuration without sliding or detachment of the contacting surfaces, friction forces are less than the normal reaction forces multiplied by the coefficient of static friction.

A necessary and sufficient condition for the workpiece stability can be obtained by setting the reaction forces within an appropriate range. Upper bound on reaction forces may be determined by the yielding strength of the contact region of the locators and lower bound may be set to be a minimum positive value with respect to safety consideration. It is therefore sensible to monitor the reaction forces throughout the machining process and make sure they fall within the range. The reaction force can be used as an indicator of the workpiece stability.

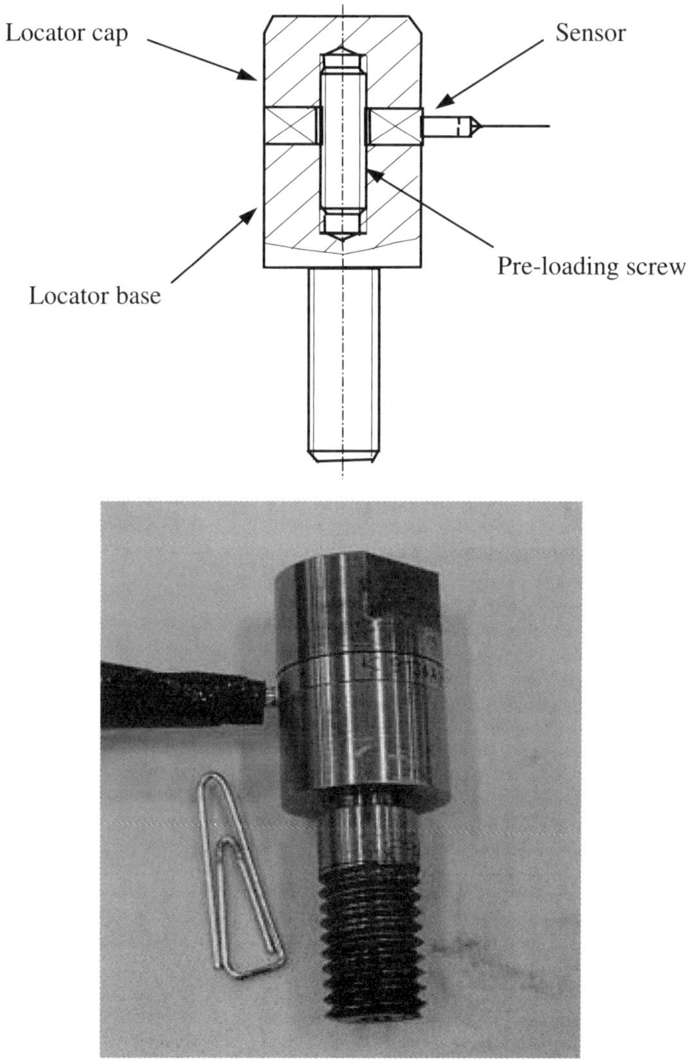

Figure 5.1 Locator embedded with piezo-electric sensor: design and prototype

5.2.1.2 Sensor Embedded Locator

In order to monitor and measure the reaction force on a locator, a specially designed locator pin is fabricated and assembled as shown in Figure 5.1. It consists of four components: a ring-shaped piezo-electric single-channel quartz sensor, a locator pin cap, a locator base and a pre-loading screw. The sensor used in this study is a Kistler SlimLine Sensor (type 9134A) and it is incorporated in the locator pin in such a way that the force to be measured acts on the sensor through the locator cap. The sensor is connected to a data acquisition system through digital interface for collecting force data.

The mounted sensor is calibrated using standard weights in order to correct any loss in sensitivity that could have occurred due to pre-loading. For maintaining the sensor sensitivity, both mounting faces of the cap and the base should be carefully ground and finished. The introduction of sensor into locator does not alter the rigidity and functionality of the fixturing element itself.

Reaction forces are dynamic and can change rapidly in magnitude. Owing to the high-sensitivity and high-rigidity characteristics of the quartz sensor, it is proven very suitable and reliable for measuring the reaction forces.

5.2.1.3 Data Acquisition for Reaction Forces

Signals from the charge amplifier connected to the sensors are continuous streams of analogue voltages and have to be sampled at a certain rate and converted into digital signals. A data acquisition board with 8 A/D differential input channels with 12-bit resolution was employed to acquire incoming data from the force sensors through the charge amplifier as shown in Figure 5.2.

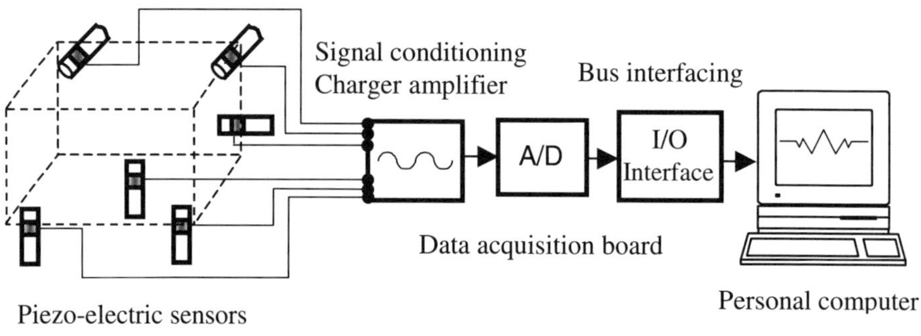

Figure 5.2 Data acquisition system for monitoring reaction forces

As there are six locators in the fixture and each of them is equipped with a sensor, multiple channel A/D conversions at defined time intervals need to be carried out. Data acquired are stored in a raw, binary format to save the time needed to convert them from the coded values to voltages.

The binary file contains a file header that includes the information necessary to convert the raw data at a later time such as the coding method used, the resolution, and the voltage range. The raw data is post-processed by a dedicated routine so that the voltage values are obtained and the digitised reaction force values are finally acquired. The flowchart of the conversion program is shown in Figure 5.3.

The acquired data are filtered using a mean-smoothing technique. In this method, the mean of a specified number of data points surrounding the data point of interest is used to calculate the smoothed data point so that noise is favourably suppressed.

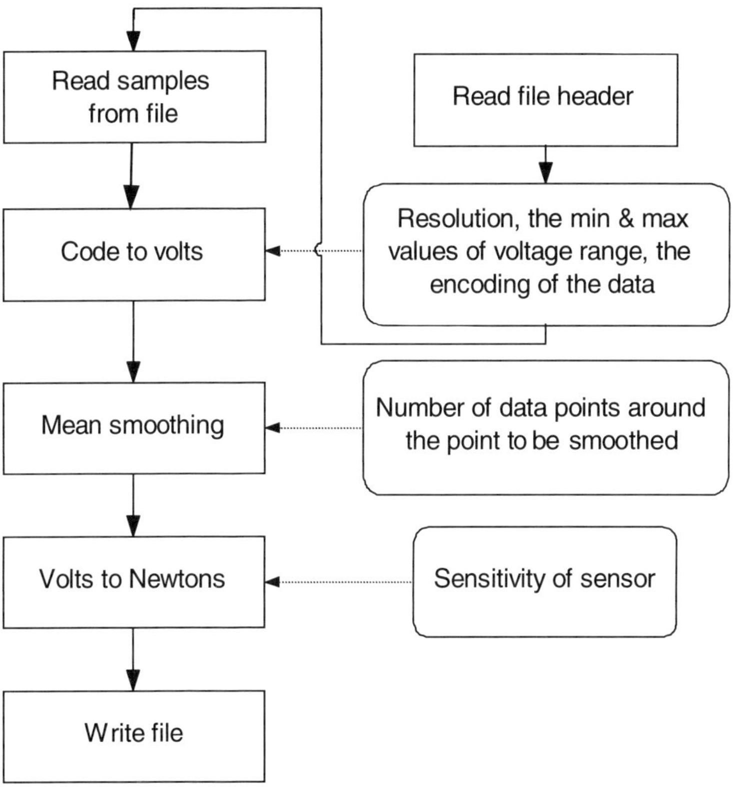

Figure 5.3 Flowchart of voltage to Newton conversion

5.2.1.4 Physical Configuration

The testing fixture is constructed using a set of modular fixturing elements including six sensor-integrated locator pins, two clamps and a baseplate as shown in Figure 5.4. The locators and clamps are arranged on the baseplate in such a way that through slots can be machined on the top face of a block-shaped workpiece by the end milling operation. The primary (base) locators are grouped in a triangular pattern on the baseplate to support the workpiece while the secondary (side) and tertiary (end) locators are positioned in a horizontal plane passing through the centre of gravity of the workpiece to restrain the remaining degrees of freedom of the workpiece.

Figure 5.4 A prototype of sensor-integrated fixture (loaded with a workpiece)

5.2.2 Experimentation

Two typical clamping schemes are investigated in the experiments in order to characterise the reaction forces on the locators during machining. First, a top downward clamping methodology is used in which strap clamps are employed to provide vertical pressure for holding the workpiece against the primary locators. The side clamping methodology is used next in which hydraulically operated clamps are driven to grasp the sides of the workpiece against the secondary and tertiary locators. For prismatic workpieces, these two workholding methods are widely used in practice.

5.2.2.1 Machining Operation

An end milling operation is performed to produce a slot on the top face of a block-shaped workpiece. The movement of the cutter is parallel to, and in the negative direction of, the *x*-axis as shown in Figure 5.5. The machining operation is performed on a Makino FNC 74-A vertical machining centre. The size of the 7075 aluminium workpiece is 220(*X*) x 122(*Y*) x 112(*Z*) mm. All the cutting operations are performed using an 18 mm diameter, coated HSS end mill with four flutes. The reaction force signals are collected by the data acquisition system with a sampling rate set at 1000 Hz. The cutting parameters used in the milling tests are listed in Table 5.1.

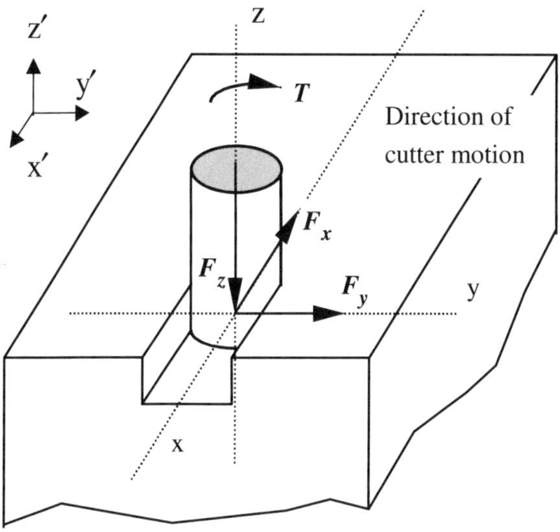

Figure 5.5 Forces and torque acting on the workpiece during end milling

Table 5.1 Cutting parameters used in the experiment

Axial depth of cutter (mm)	Feed per tooth (mm)	Spindle speed (rpm)	Feed rate (mm/min)	Coolant
3	0.05	500	100	air

According to the cutter geometry, cutting conditions and workpiece properties, the milling forces can be easily generated using the cutting force model (Kline, 1982). The moment of milling (torque) is deduced from the power consumption in terms of

$$\boldsymbol{T} = 5306.4\frac{\boldsymbol{P}}{\boldsymbol{n}} \tag{5.1}$$

where $\boldsymbol{T}$ is the torque (Nm), $\boldsymbol{P}$ is the power consumption (kW) and $\boldsymbol{n}$ is the spindle speed (rpm). The estimated milling forces and torque are computed and tabulated in Table 5.2.

Table 5.2 Estimated milling force components and torque

Average F_x (N)	Average F_y (N)	Average F_z (N)	Torque T (Nm)
-131	232	-55	2.77

5.2.2.2 *Top Clamping*

The fixture layout with top clamps is shown in Figure 5.6. Clamping forces are applied through tightening the clamp nut with a torque wrench. The clamping forces applied on the workpiece are determined using the static equilibrium equations and the reaction forces on the base locators (nos. 0, 1, 2). The values of the applied forces are tabulated in Table 5.3.

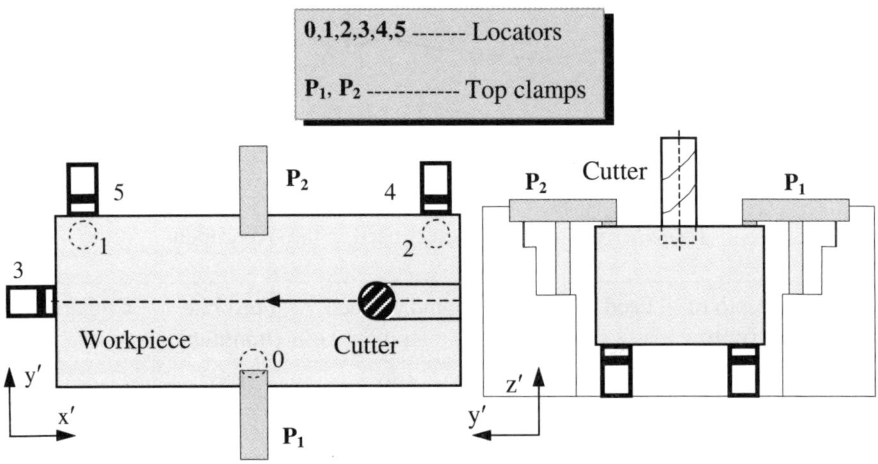

Figure 5.6 Layout of fixture with top clamps

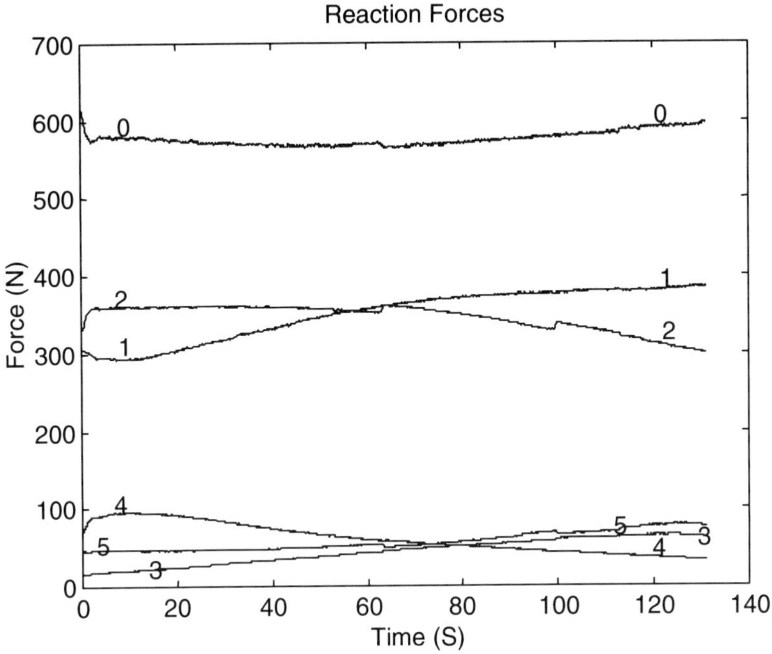

Figure 5.7 Measured reaction forces in the case of top clamping

The measured reaction forces are shown in Figure 5.7. Detailed discussions on the obtained results are presented in Section 5.2.3.1.

Table 5.3 Clamping forces used in the top clamping experiments

Clamp No.1 P_1 (N)	Clamp No.2 P_2 (N)
690	610

5.2.2.3 *Side Clamping*

The fixture layout with side clamps is shown in Figure 5.8. The clamping forces are hydraulically applied by activating a hand pump. Clamping forces are calculated using the pressure readings from fluid pressure gauges and are listed in Table 5.4. The measured reaction forces are shown in Figure 5.9. Detailed discussions on the obtained results are presented in Section 5.2.3.2.

Table 5.4 Clamping forces used in the side clamping experiments

Clamp No.1 P_1 (N)	Clamp No.2 P_2 (N)
640	670

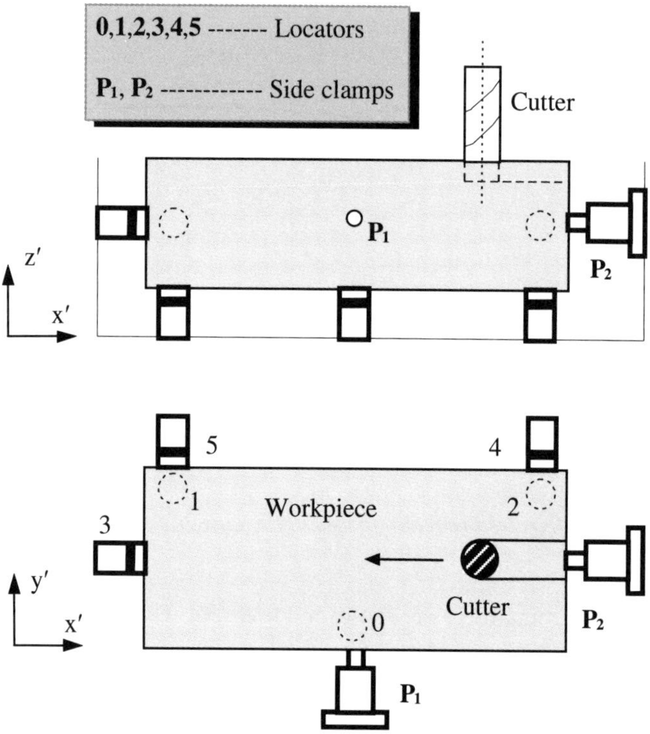

Figure 5.8 Layout of fixture with side clamping

5.2.3 Discussion on Reaction Forces

It is interesting to note that there is a large difference in magnitudes of reaction forces between the primary, secondary and tertiary locators for both top and side clamping. In the case of top clamping as in Figure 5.7, the reaction forces on locator nos. 0, 1 and 2 are much larger than those on locator nos. 3, 4 and 5. It indicates that locators no. 0, 1 and 2 play a bigger role in counteracting the cutting forces, thus holding the workpiece in its position. Most of the clamping forces are balanced by locators no. 0, 1, 2, which also implies that they are subjected to higher pressure during the machining process and are more likely to cause workpiece displacement. On the other hand, it is clear that the workpiece is prone to dislodge from locators no. 3, 4, 5.

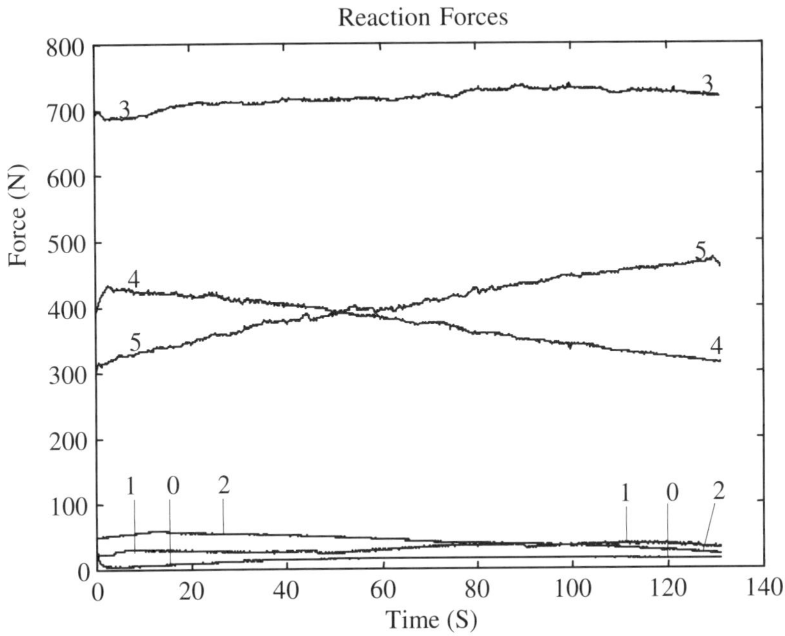

Figure 5.9 Measured reaction forces in the case of side clamping

On the contrary, when the workpiece is held by side clamping as in Figure 5.9, the reaction forces are opposite to the top clamping case, with the secondary and tertiary locators now playing a more significant role than the primary ones.

Therefore, those locators with substantially higher reaction forces may be defined as *positive* locators and those with considerably lower reaction forces as *subordinate* locators according to their performance in fully supporting the workpiece. In the case of top clamping, the positive locators are the base ones and the subordinate locators are the side and end locators. Similarly in the case of side clamping, the positive locators are the side and end ones and the subordinate locators are the base locators.

When it comes to preventing the locator from yielding under pressure or to prohibit large local deformation at the contacting points, the reaction forces on the positive locator should be used to evaluate the strength of the locator or the local rigidity of the workpiece. All compressive stresses should be entirely within the strength limit of the locators. On the other hand, for prohibiting the workpiece from displacement under the influence of cutting forces and vibration, the reaction forces on the subordinate locators should be constantly and closely monitored. Applied clamping forces should be ideally sufficient to hold the workpiece against the locators. Excessive clamping forces must be avoided as they may damage the workpiece or the locators. It is preferred to use the primary instead of the secondary or tertiary locators as positive locators to resist against the cutting and clamping forces.

5.2.3.1 Top Clamping

In the case of top clamping (Figure 5.7), the following observations can be made from the experimental results:

- When the cutter moves in the negative direction of x-axis, the reaction force on locator No.1 (curve 1) steadily increases while that on locator No.2 (curve 2) gradually decreases, and their variances are almost the same.

- Curves 1 and 2 intersect with each other when the cutter comes in the vicinity of the centre of gravity of the workpiece, which can be intuitively expected from the static equilibrium condition.

- Curve 0 is quite stable with very small variation while the lowest reaction force occurs around the centre of gravity of the workpiece.

- Curve 3 slightly increases and it reaches its peak (about 50N) when the cutter exits its toolpath. It can be inferred that the x-force component of the cutter is primarily counteracted by the friction between the workpiece and the fixturing elements, rather than by the end locator alone.

- Throughout the milling operation, the reaction force on locator No.5 steadily increases while that on locator No.4 constantly decreases, both with modest fluctuation.

5.2.3.2 Side Clamping

In the case of side clamping as in Figure 5.9, the following observations can be made from the experimental results:

- Reaction force on locator No. 3 is almost constant with very small variation. This is because the reaction force at locator No. 3 is approximately the addition of constant force $\boldsymbol{P}_2$ and the x-directional component of the milling force. The fluctuation of the x-axis component of the milling force is very small compared to the clamping forces.

- When the cutter moves along its tool-path, the reaction force on locator No.5 steadily increases while that on locator No.4 constantly decreases. This is due to the y-directional component of the cutting force shifting its point of application during machining.

- Curves 4 and 5 intersect with each other when the cutter comes in the vicinity of the centre of gravity of the workpiece, which can be perceptively expected from the equilibrium condition.

- Curve 0 is very close to zero. It can be interpreted to mean that the base locator can be out of contact easily as the workpiece is prone to tilt slightly when it is held by its sides. It indicates that a prismatic workpiece is better held by top clamps than by side clamps.

- Curves 1 and 2 almost remain constant throughout the entire process. Although they are quite low in values, they never drop to zero, which means that they are constantly in contact with the workpiece and the workholding is stable.

5.2.3.3 Spectral Analysis of Reaction Forces

Reaction forces are inherent dynamic variables, and it is reasonable to examine them both in time and frequency domains. In previous section, the dynamics of reaction forces is observed and evaluated from the point of view of temporal changes. It is now focused on the dynamic variation of the reaction forces with the help of spectral analysis, therefore, linear trends such as the components contributed by the fixed clamping forces are removed from the sampled data by giving them a zero arithmetic mean, so that spectral analysis are carried out using the Fast Fourier Transform (FFT).

For comparison, spectral analysis results of the reaction forces on locators no. 0 and 3 with top clamping are presented. The measured raw reaction forces are shown in Figures 5.10(a) and 5.11(a) respectively, in which a snapshot containing 0.12 second (corresponding to one revolution of the cutter) worth of data with 120 points of values for each locator is depicted. However, for the power spectrum analysis, 0.36 second (corresponding to three revolutions of the cutter) worth of data with 360 points of values are used. Their spectral analysis results are shown in Figures 5.10(b) and 5.11(b) respectively.

As the prominent component of the cutting force transients, tooth passing frequency can be represented as follows:

$$f_n = \frac{nZ}{60} \tag{5.2}$$

where f_n is the tooth passing frequency (Hz), $\boldsymbol{n}$ is the spindle speed (rpm) and $\mathbf{Z}$ is the tooth number of the cutter. Therefore, the primary frequency of the milling forces is calculated to be 33.3 Hz by considering the spindle speed $\boldsymbol{n}$ = 500 rpm and the tooth number $\mathbf{Z}$ = 4.

It is obvious from Figure 5.10(a) that there are four distinctive fluctuations of the reaction force within the time frame of one rotation of the cutter, which concurs with the fact that milling operation is an intermittent cutting process and the end mill has four flutes. The peak-valley variation of the reaction force is about 110N. In Figure 5.11(a), however, the changes are not distinct and the overall trend is flat. Its variation range is only 20N, which is much smaller than the former. It could mean that the reaction force on locator No.0 is significantly influenced by the dynamic alternation of the milling force while locator No.3 is much less affected.

For the dynamic variation of the reaction forces on locator No.0, its most important component is contributed by the four-flute cutter as shown in Figure 5.10(b), and the second largest is from the vibration due to cutter runout. On the other hand, as shown in Figure 5.11(b), for locator No.3, the influence of vibration from cutter run-out is larger than the cutter. Therefore, it indicates that the dynamic change of the reaction forces on the positive locators is more affected by the cutting force fluctuation than the subordinate locators, implying that a positive locator interacts more closely with the workpiece and it should be used to monitor the dynamic performance of the locators.

5.3 Finite Element Analysis

In this section, the development of a finite element methodology to study the behaviour of surface interaction between the workpiece and the fixturing elements during machining is explained. The non-linear surface contacts and frictional effects are modelled with the finite element method. FEM modelling of the workpiece-fixture system is developed and is experimentally verified by comparing the computational results with the measured reaction forces. This approach is useful in optimal design and verification of fixture configuration and provides guidelines for constructing an intelligent fixture.

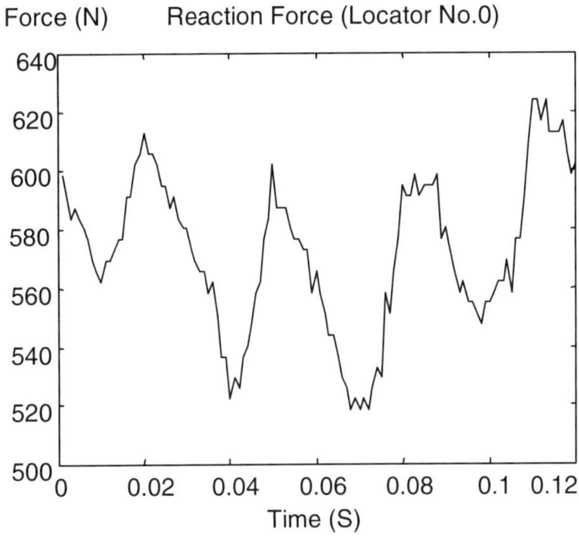

(a)

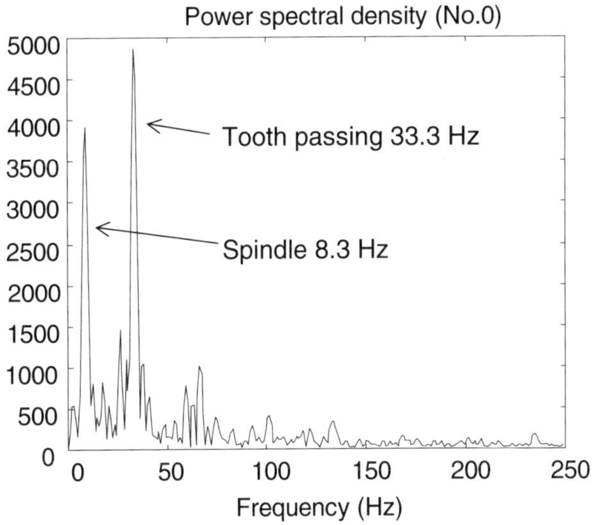

(b)

Figure 5.10 (a) Measured reaction force on locator No. 0 and (b) power spectrum (top clamping)

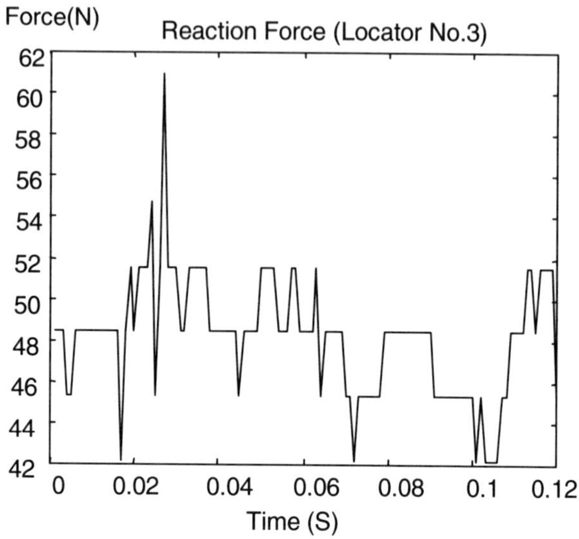

(a)

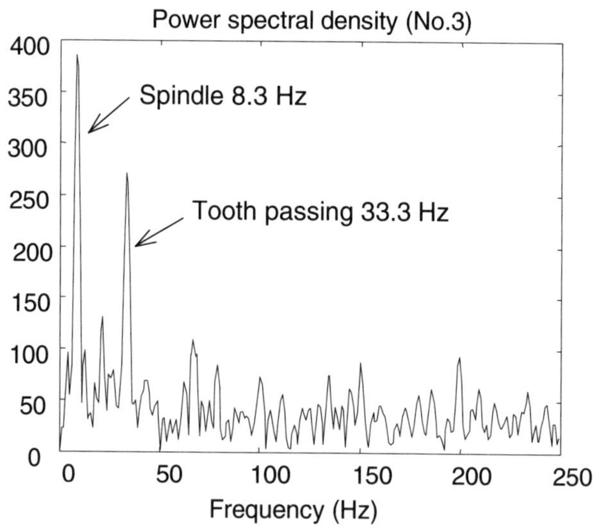

(b)

Figure 5.11 (a) Measured reaction force on locator No.3 and (b) power spectrum (top clamping)

5.3.1 Contact Force/Deflection Relationships

A contact pair consists of two components that can be arbitrarily defined as a contactor and a target (these terms may be replaced by slave and master respectively without ambiguity). Points on the contactor surface are inhibited from penetrating the target and only allowed to slide on the target tangent plane. For the ease of analysis, a reference frame with unit vectors $\boldsymbol{n}$, and $\boldsymbol{t}_1$ and $\boldsymbol{t}_2$ is defined at discrete contact points on the target surface (Figure 5.12). The outward normal to the plane is defined by $\boldsymbol{n}$, and $\boldsymbol{t}_1$ and $\boldsymbol{t}_2$ are mutually perpendicular tangents. The normal gap $\boldsymbol{g}$ is defined as positive, and the contact force as negative with respect to $\boldsymbol{n}$. Sliding deflections and friction forces can be expressed as a linear combination of $\boldsymbol{t}_1$ and $\boldsymbol{t}_2$. The direction of friction force is opposite to the relative motion.

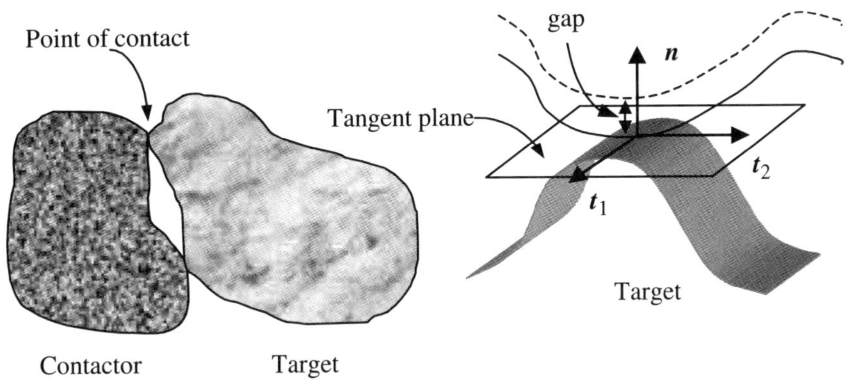

Figure 5.12 Contacting bodies as contactor and target

The interaction between contacting bodies can be described by a complementarity expression called the Kuhn-Tucker condition (Yagawa and Kanto, 1993). In finite element analysis, it is convenient to apply these conditions with regard to discrete contacts. Therefore, the state of contact can be represented as follows:

1. for normal contact

separation	$\boldsymbol{g} \geq 0$	
sticking	$\boldsymbol{c} \leq 0$	(5.3)
mutual exclusivity	$\boldsymbol{gc} = 0$	

where ***g*** is the normal clearance (gap) measured from the target surface and ***c*** is the normal force acting upon the target surface at the point of contact. This complementary relationship can be clearly illustrated in Figure 5.13.

2. for sliding contact

$$\begin{array}{ll} \text{sticking} & \phi \leq 0 \\ \text{sliding} & \gamma \geq 0 \\ \text{mutual exclusivity} & \phi\gamma = 0 \end{array} \tag{5.4}$$

where ϕ is the frictional capacity minus the tangential component of the contact force, and γ is the absolute value of the relative sliding velocity between the contacting points.

The implementation of the contact and sliding conditions can be considered as a problem of constrained minimisation. Lagrange multiplier and penalty approaches can be used to solve the problem (Simo and Laursen, 1992).

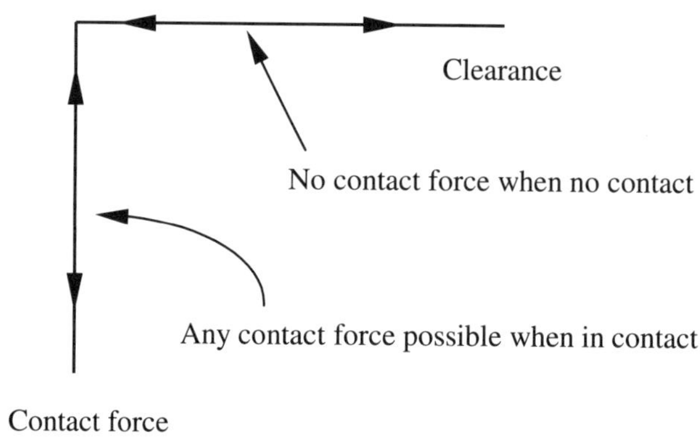

Figure 5.13 The relationship between the contact force and the gap

5.3.2 Contact Approach

One of the approaches to represent the contact problem is to use gap elements with node-to-node connections between the contact pair. The basic variables involved here are the clearance, the normal direction and the force transmitted. The contact direction and load path are defined according to the initial mesh configuration. The normal vector is defined such that it passes through the two nodes on the contact pair. A small sliding is allowed to occur along a tangent perpendicular to the normal. For each potential contact point, the contact condition is described by the clearance as follows:

$$g = g_0 + \boldsymbol{n} \cdot (\boldsymbol{u}_2 - \boldsymbol{u}_1) \tag{5.5}$$

where g_0 is the initial clearance, $\boldsymbol{u}_1$ and $\boldsymbol{u}_2$ are displacements of the corresponding gap element nodes. The contact state of sticking, sliding and release is determined by using nodal contact forces.

Another approach is node-to-surface approach that can appear in the guise of gap elements or slide-line elements. The nodes on the contactor body (slave) are prevented from penetrating the surface of the target body (master), and the contact direction is always kept normal to the master surface. The tangential surface tractions are coupled to the normal contact pressure through frictional effects. This is a more general method than the gap element, in which finite rigid body sliding can be accommodated by redefining the normal and tangent vectors using updated geometry data.

5.3.3 FEM Model and Solution

In this section, the workpiece is modelled as a deformable body based on linear elasticity, and isotropic Coulomb friction is included. Contact elements are used to simulate the contact phenomena between the workpiece and the fixturing elements. The fixture elements can only exert compression on the workpiece, otherwise, the nodal point in question should be regarded as being detached from the fixture elements.

FEM analysis is performed to simulate the workpiece-fixture system during the milling process so as to compute the reaction forces on each locator, deformation of the workpiece and the distribution of stresses on the contact elements. The complete analysis is accomplished using a non-linear finite element code ABAQUS (Hibbitt, 1995) on a DEC Alpha workstation.

5.3.4 Model Validation

For model verification, the experimental tests in Section 5.2 are simulated by the FEM approach. The aluminium block-shaped workpiece is modelled and meshed using 8-noded brick elements as shown in Figure 5.14 and the quasi-static analyses of the model due to cutting force, location of the cutter, position of the fixturing elements are carried out when a through slot is milled. The cutting forces and the torque acting on the workpiece are the same as shown in Table 5.2. The fixturing elements are modelled using a 2D plate element while the contact surface between the fixturing element and the workpiece is modelled with a frictional coefficient of 0.2. The Young's modulus, $\boldsymbol{E}$, and the Poisson's ratio, $\boldsymbol{\nu}$, of the workpiece are set to be 6.89E4 MPa and 0.33 respectively. The contact surface on the workpiece is treated as a slave surface while the fixturing element surface is treated as the master surface. This is to ensure that deformation occurs mainly on the workpiece.

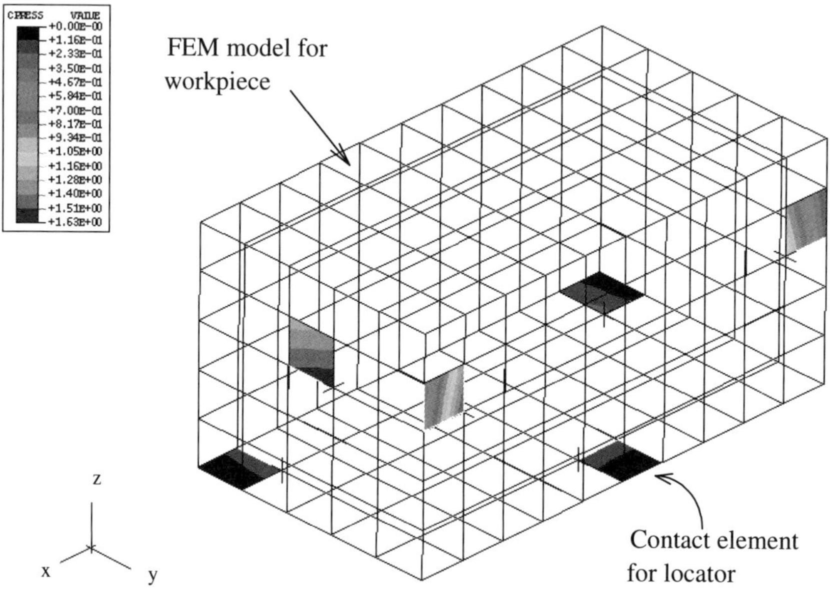

Figure 5.14 FEM model of the block workpiece

5.3.4.1 Comparison of Reaction Forces

The reaction forces at the contacts between the workpiece and the fixture for both top and side clamping are generated by FEM analysis and superimposed with the experimentally measured reaction forces as shown in Figures 5.15 and 5.16 respectively. It is clear from the figures that the calculations agree reasonably well with the experimental values, for both cases of top and side clamping. The fluctuating trends of the reaction forces and the magnitudes of the forces from the FEM calculations and the experiment are comparable. Thus, the FEM approach has been validated and can be further extended to study the deformation of workpiece under dynamic clamping.

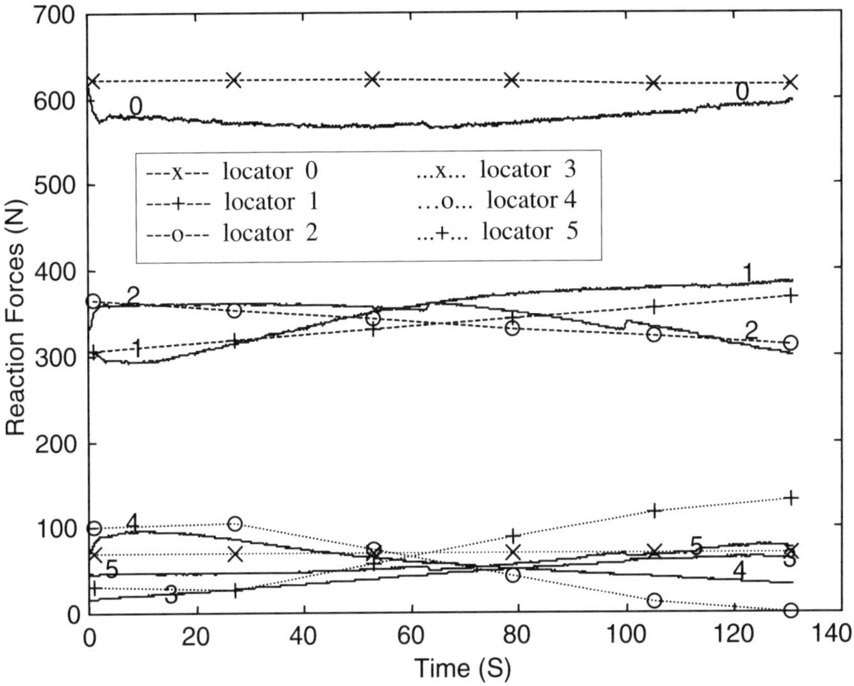

Figure 5.15 The measured reaction forces superimposed with calculated values (dotted and dashed) for top clamping case

5.3.5 Deformation under Dynamic Clamping

One of the fixturing requirements is that the fixture must limit the deformation of the workpiece. It is important to consider both the cutting forces as well as the clamping forces on the workpiece, which cause the workpiece deflection and deformation depending on the elasticity of the workpiece. Workpiece distortion is always a concern in machining accuracy control, especially for delicate workpieces like thin-walled components.

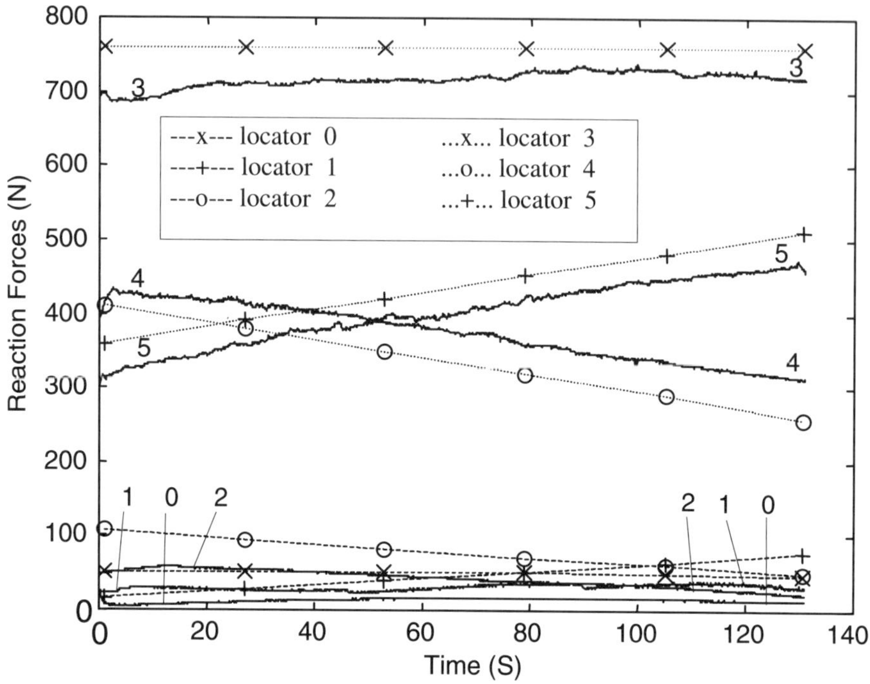

Figure 5.16 The measured reaction forces superimposed with calculated values (dotted and dashed) for side clamping case

The elastic deformation of a workpiece induced by the clamping and cutting forces can be efficiently estimated using FEM analysis. When a workpiece is being machined, the possibility of deformation is very high, which is a function of the material properties, geometry of the workpiece, clamping and cutting forces. When a machined workpiece is released from the clamps, the earlier distortion may reappear (completely elastic) but with the converse sign. Note that not all the distortions throughout the entire workpiece may contribute to the dimensional errors of a machined surface. Shape deformation within the characteristic radius of the cutter contacting point may give rise to machining errors.

Dynamic clamping forces refer to the action of changing the clamping intensities as the cutter moves into the different positions of the workpiece. The clamping intensities are optimally computed considering the instantaneous machining and the external forces as in Chapter 4. In addition, another dynamic clamping scheme with dynamically varying clamping points with respect to the cutting dynamics is also investigated, although on-the-fly shift of clamping position is not easy to implement for the moment.

In order to accentuate the effect of dynamic clamping on workpiece distortion, a hollow workpiece model (Figure 5.17) is examined using the FEM analysis, which is subjected to the same loading and boundary conditions as the solid workpiece model in Figure 5.14. The hollow workpiece is located by six locators according to the 3-2-1 principle and held by two downward clamps applied at the top surface. For ease of comparison, the workpiece deformation under conventional constant clamping is first analysed. The deflections around the area of interest throughout the entire tool path in *x*, *y* and *z* directions are shown in Figure 5.18.

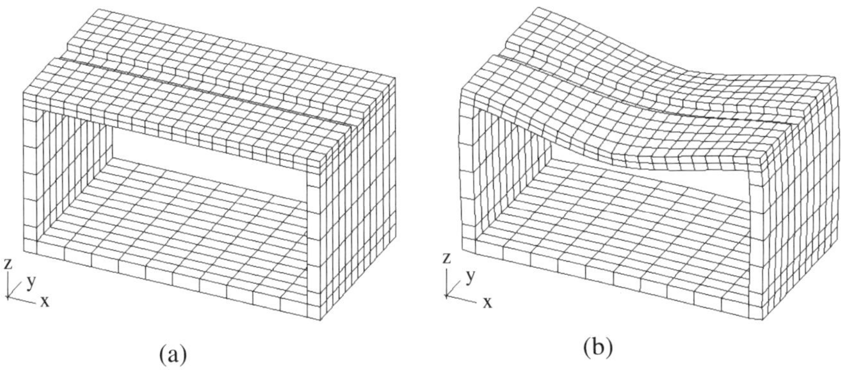

Figure 5.17 (a) FEM model of a hollow workpiece and (b) typical deformed shape of the workpiece

As shown in Figure 5.18, *z*-directional deflection δ_Z is much larger than those at *x*- or *y*-direction which can be ignored in error analysis. The variation of δ_Z would give rise to the deviation of slot depth. All the deflection is contributed by both cutting and clamping forces. The direct influence of the cutting forces cannot be eliminated as they are always required for the machining operation. Therefore, it is natural to reduce the deflection induced by clamping through manipulating the clamping forces in both magnitudes and positions.

For a clear comparison, the deflections δ_Z under constant and dynamic clamping are shown together in Figure 5.19. The deflection decrements due to dynamic clamping conditions are found to be 34.9% and 48.4% respectively. In the dynamic clamping position strategy, the two top clamps are shifted opposite to each other as the cutter moves during machining. An abrupt drop in deflection as the cutter approaches the plane containing the centre of gravity can be observed. It can be concluded that a dynamic clamping scheme can effectively reduce workpiece deformation, and this is beneficial to the enhancement of the finished accuracy of the workpiece.

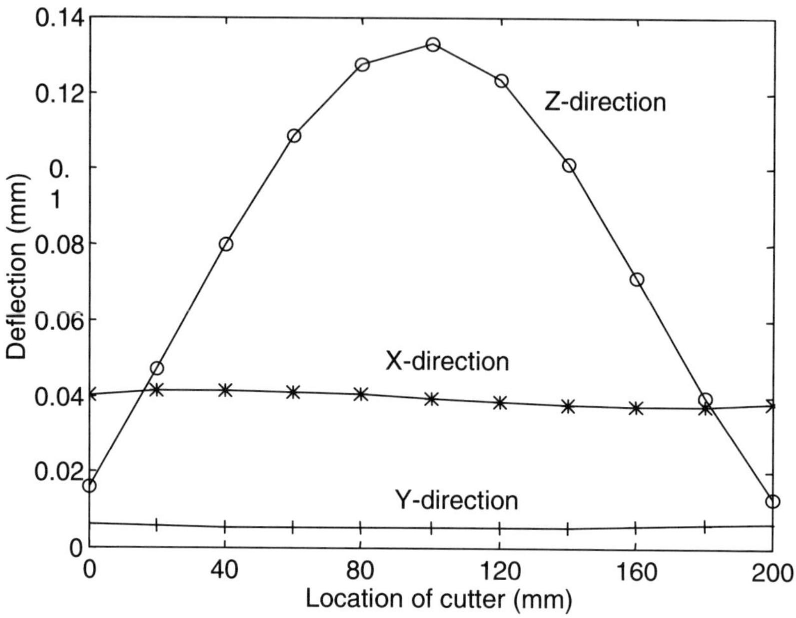

Figure 5.18 Deflection along cutter path under constant clamping

5.4 Summary

The experimental investigation of workpiece-fixture interaction in terms of dynamic reaction forces on locators is presented. The measured reaction forces represent the actual interaction between the interfaces of the workpiece and the fixturing elements during machining. They are good indicators of the workpiece stability. It is experimentally demonstrated that friction does play an important role in workholding. The experimental data provides insight into the dynamic nature of the workpiece-fixture system.

The developed FEM model provides an excellent tool to analyse the fixture configurations as well as the effect of the fixturing parameters on workpiece deformation before the actual workholding execution. Good agreement is found between the actual interaction forces and the simulated results. The validated model is employed to predict the workpiece deformation under the dynamic clamping schemes. It is shown that approach to enhance workpiece accuracy through the manipulation of fixturing parameters in terms of clamping dynamics is promising.

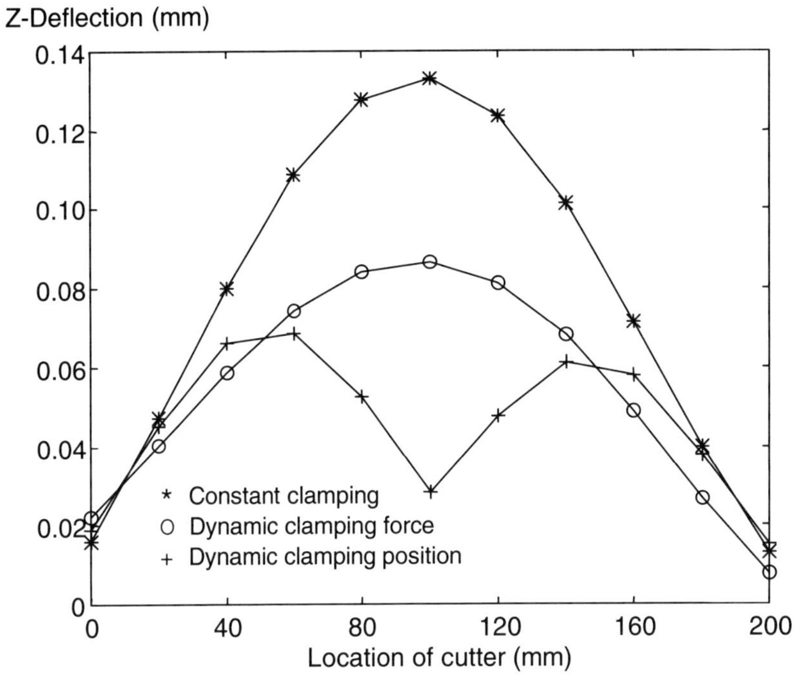

Figure 5.19 Workpiece deflection under constant and dynamic clamping

5.5 Bibliography

Asada, H. and A.B. By. Kinematic Analysis of Workpart Fixturing for Flexible Assembly with Automatically Reconfigurable Fixtures. *IEEE Journal of Robotics and Automation*, 1(2), pp. 86-94. 1985.

Chou, Y.-C., V. Chandru and M.M. Barash. A Mathematical Approach to Automatic Configuration of Machining Fixtures: Analysis and Synthesis. *Journal of Engineering for Industry*, 111(4), pp.579-584. 1989.

Cogun, C. The Importance of the Application Sequence of Clamping Forces on Workpiece Accuracy. *Journal of Engineering for Industry*, 114, pp.539-543. 1992.

DeMeter, E.C., Q. Sayeed, R.E. DeVor and S.G. Kapoor. An Internet-based model for technology integration and access, Part 2: Application to process modelling and fixture design. *Symposium on Agile Manufacturing, ASME IMECE*, 1995.

Hibbitt, Karlsson and Sorensen, Inc. *ABAQUS User's Manual*. 1995.

Hockenberger, M.J. and E.C. DeMeter. The Effect of Machining Fixture Design Parameters on Workpiece Displacement. *Manufacturing Review*, 8(1), pp.22-32. 1994.

Kline, W.A. *The Prediction of Cutting Forces and Surface Accuracy for the End Milling Process*. PhD Thesis, University of Illinois at Urbana-Champaign. 1982.

Lee, J.D. and L.S. Haynes. Finite-Element Analysis of Flexible Fixturing System. *Journal of Engineering for Industry*, 109(2), pp.134-139. 1987.

Lee, S.H. and M.R. Cutkosky. Fixturing Planning with Friction. *Journal of Engineering for Industry*, 113, pp. 320-327. 1991.

Liao, Y. and S.J. Hu. An Integrated Model of Fixture-workpiece System for Surface Quality Prediction. *Int J. Adv. Mfg. Technol.*, Vol.17 (11), pp.810-818. 2001.

Liao, Y.J. and S.J. Hu. Flexible Multibody Dynamics Based Fixture-Workpiece Analysis Model for Fixturing Stability. *International Journal of Machine Tools and Manufacture*, Vol.40, No.3, pp.343-362. 2000.

Mittal, R.O., P.H. Cohen and B.J. Gilmore. Dynamic Modelling of the Fixture-Workpiece System. *Robotics and Computer-integrated Manufacturing*, 8(4), pp.201-217. 1991.

Nee, A.Y.C. and A. Senthil kumar. Workpiece Accuracy and Tolerance Analysis in Modular Fixtures. In *Advanced Tolerancing Techniques*, ed. by H.C. Zhang, pp. 355-379. John Wiley & Sons. 1997.

Pham, D.T. and A. de Sam Lazaro. Finite Element Study of a Workpiece in a Machining Fixture. *Mathematical and Computer Modelling*, Vol.14, pp.1024-1028. 1990.

Raghu, A. and S.N. Melkote. Analysis of the Effects of Fixture Clamping Sequence on Part Location Errors. *International Journal of Machine Tools and Manufacture*, Vol.44, No.4, pp.373-382. 2004.

Rong, Y., W. Hu, Y. Kang, Y. Zhang and D.W. Yen. Locating Error Analysis and Tolerance Assignment for Computer-aided Fixture Design. *Int. J. Prod. Res.*, Vol. 39(15), pp.3529-3545. 2001.

Shawki, G.S.A. and M.M. Abdel-Aal. Effect of Fixture Rigidity and Wear on Dimensional Accuracy. *International Journal of Machine Tool Design and Research*, 5, pp.183-202. 1965.

Shawki, G.S.A. and M.M. Abdel-Aal. Rigidity Considerations in Fixture Design - Contact Rigidity at Locating Elements. *International Journal of Machine Tool Design and Research*, 6, pp.31-43. 1966.

Shawki, G.S.A. and M.M. Abdel-Aal. Rigidity Considerations in Fixture Design - Contact Rigidity of Clamping Elements. *International Journal of Machine Tool Design and Research*, 6, pp.207-220. 1966.

Shawki, G.S.A. Rigidity Considerations in Fixture Design - Contact Rigidity for Eccentric Clamping. *International Journal of Machine Tool Design and Research*, 7, pp.195-205. 1967.

Simo, J.C. and T.A. Laursen. Augmented Lagrangian Treatment of Contact Problems Involving Friction. *Computers & Structures*, 42(1), pp.97-116. 1992.

Trappey, A.J.C., C.S. Su and J.L. Hou. Computer-Aided Fixture Analysis Using Finite Element Analysis and Mathematical Optimisation Modelling. *Manufacturing Science and Engineering ASME*, Vol. 3-1, pp.777-787. 1995.

Yagawa, G. and Y. Kanto. Finite Element Analysis of Contact Problems Using Penalty Function Method. In *Computational Methods in Contact Mechanics*, ed. by M.H. Aliabadi and C.A. Brebbia, pp. 127-154. Elsevier Applied Science. 1993.

Yu, K.M., T.W. Lam and A.H.C. Lee. Immobilization Check for Fixture Design. *Proc. Instn. Mech. Engrs., Part B: J. Engineering Manufacture*, Vol.217, pp.499-512. 2003.

Zhang, Y., W. Hu, Y. Rong and D.W. Yen. Graph-based Set-up Planning and Tolerance Decomposition for Computer-aided Fixture Design. *Int. J. Prod. Res.*, Vol.39(14), pp.3109-3126. 2001.

Chapter 6

An Intelligent Fixturing System

6.1 Introduction

Both proper fixture design and optimum fixturing execution are crucial to workpiece quality assurance during machining, assembly or inspection. In the context of computer-integrated manufacturing, much research effort has been focused on the fixture analysis and planning such as the kinematics analysis, generation of viable fixturing configuration, workpiece deformation analysis, AI application in fixture design, etc. Whereas fixture design methodology has been well investigated, little work has been directed towards the equally critical task of fixturing execution. A missing thread in the past research activities is to investigate the performance of a fixture during the machining stage and how to guarantee and improve the machining accuracy through concurrent control of workholding operation with machining process. Therefore, a better approach to the fixturing problem is to integrate optimal fixture design with proper fixturing execution in a holistic procedure.

This chapter deals with the architecture of an intelligent fixturing system, which incorporates a fixturing planner that generates viable fixture configuration as well as a corresponding "live" fixture that can correct workpiece position through performing tool path compensation before machining. It can also adaptively vary clamping forces with respect to dynamic cutting forces so as to minimise the workpiece deformation during machining. The system configuration is first introduced, followed by tool path compensation techniques. The need for dynamic clamping is then justified from the point of view of workpiece quality. Being an important part of the "live" fixture, a novel dynamic clamping actuator capable of delivering time-varying clamping intensities is presented. In order to adaptively control the clamping forces, a simplified generalised minimum variance (GMV) approach is proposed and discussed. Finally, the impact of the intelligent fixturing system on accuracy, surface quality and cycle time is demonstrated through a series of experimental work.

6.2 Architecture of an Intelligent Fixturing System

6.2.1 System Functions: From Viable Design to Optimal Execution

The metric of fixture performance is always product quality. A good fixture design is one that maximises geometric accuracy and surface quality of a finished workpiece. The fixturing execution is of likewise importance, playing a significant role in the minimisation of workpiece displacement and distortion during machining. With regard to the quality measure, an intelligent fixturing system is developed which incorporates three interdependent sub-functions: analysis and synthesis, fixture hardware generation and workholding execution (Figure 6.1). Each sub-function contributes to the final quality of the finished workpiece and should be treated with equal attention. A discerning implementation of workholding operation starts with an analytical fixture planning procedure which would generate a fixture design that can accurately locate and hold the workpiece with respect to the machine tool and a set of optimised fixturing parameters.

The next sub-function is the assembly of a "live" fixture hardware using a modular fixture kit as well as sensing and control devices, clamping drivers, electronics and interfacing circuitry, and a personal computer. This fixture with sensors and actuators is capable of on-line control of fixturing action, as opposed to a conventional "dummy" fixture with hard-wired clamping devices. The assembly activity is preferably performed by the toolmaker and/or tool designer who, in view of the fixturing control, should have interdisciplinary knowledge such as mechatronics, computer control and digital signal processing (DSP). The output from the sub-function is an intelligent fixture assembly ready for loading workpiece for a machining operation.

The final sub-function is the execution of the adaptive workholding operation. It is critical to secure perpendicularity and parallelism of the workpiece within the fixture after clamping actuation. Therefore a probing procedure is performed to detect the genuine position and orientation of the workpiece within the fixture and correspondingly, the tool path is compensated through NC code manipulation according to the evaluated workpiece displacement. Likewise, the fixture should minimise workpiece distortion during machining. This is achieved through both providing appropriate supports and applying dynamic clamping forces. Supports should be used to uphold flexible features on the workpiece whenever possible to avoid large deflection or permanent damage to the workpiece. Just sufficient dynamic clamping forces are applied to sustain the load of the cutting force.

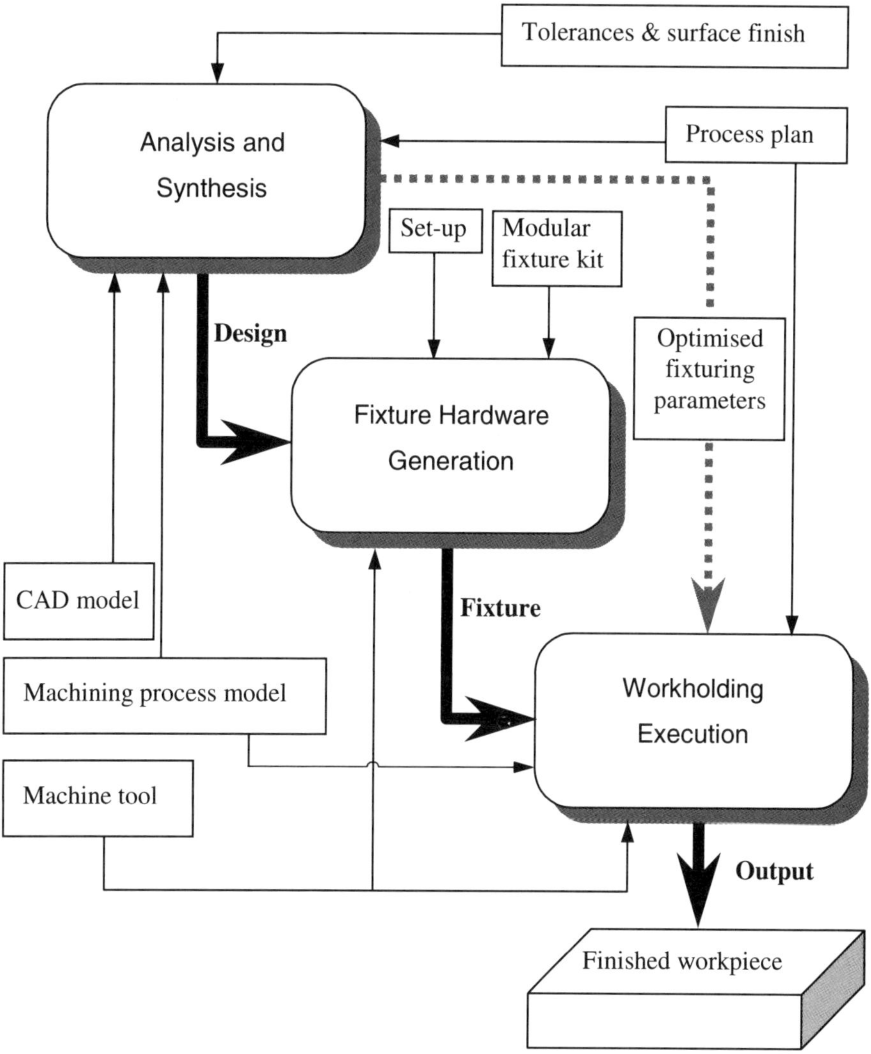

Figure 6.1 Framework of an intelligent fixturing system

6.2.1.1 Optimal Design Procedure

During the fixture design stage, a series of analyses are performed in order to generate an optimal design. As depicted in Figure 6.2, the complete procedure consists of modules such as kinematic restraint analysis, total restraint analysis, clamping intensity analysis and FEM analysis. These four modules, with progressively more realistic physical models, have been covered in detail in terms of theoretical techniques in the previous chapters. Here, the integrated design procedures with its functions are presented as follows:

- Kinematic restraint analysis

Assuming rigid body and point contact, kinematic restraint analysis is to determine a locating scheme that is sufficient to provide a uniquely accurate position for the workpiece. It can be accomplished either by an expert system based on heuristic knowledge and the experience as well as intuition of tool designers, or by an analytical system relying on twist and wrench vectors from the screw theory (Chou, 1989; DeMeter, 1994). Once the locating points and placements are determined, the type of locators and their geometry can be selected according to the availability of modular fixtures and the type of the machine tool. In this chapter, the locating layout is decided by heuristic rules such as the 3-2-1 principle to ensure proper locating specifications.

- Total Restraint Analysis

Based on the computational geometry theory and the geometric reasoning technique, this module is to automatically generate a clamping scheme and clamping sequence that is sufficient to totally constrain the workpiece. This module would also provide feedback on the adequacy of the kinematic restraint by the locating layout. At this stage, the clamps are still assumed to make point contacts with a rigid workpiece. Once the contact geometry of the clamps is designed, the resultant contact would be either point, linear or polygonal depending on the workpiece and clamp geometry.

- Clamping Intensity Analysis

Generation of clamping intensities is based on the assumption of rigid body and polygonal contacts. Quasi-static equilibrium conditions are used in calculations so that the impact of cutting force dynamics is substantially accounted for. Both minimum and maximum clamping intensities are generated by the module.

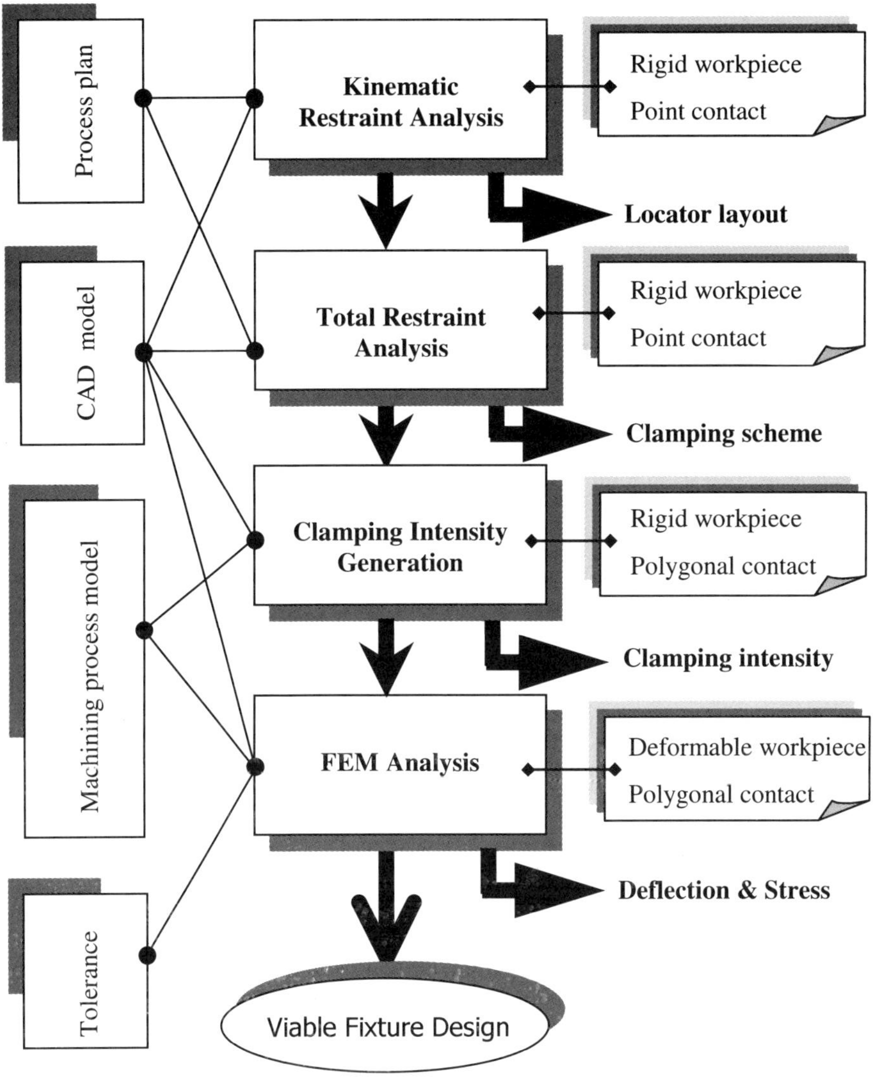

Figure 6.2 Analysis and synthesis procedure and software integration

- FEM Analysis

Workpiece displacement and distortion are due to a number of physical phenomena such as workpiece elasticity, fixture flexibility, and deflection/lift-off at the workpiece-fixture contact regions. It is usually quite severe during the machining of thin-walled features on otherwise structurally rigid workpieces. Assuming deformable workpiece and polygonal contacts, FEM analysis is performed to predict workpiece deflection and/or lift-off that would introduce form error into the finished workpiece in the pattern of lost flatness or cylindricity.

6.2.2 "Live" Fixture for Precision Machining

6.2.2.1 Clamping Sub-tasks

In the clamp planning and synthesis procedure, a clamping operation is divided into two stages: clamping actuation before machining and clamping conservation during machining. The separation of the clamping task has previously been proven helpful for generating viable clamping configuration. Nevertheless, it is also beneficial to the workholding execution as it has the advantage of facilitating the management of workpiece displacement and distortion. This leads to performance enhancement of the automated clamping task.

It is possible that tool path can be compensated to fit the actual workpiece position in a fixture (Hazen and Wright, 1988). Therefore workpiece displacement due to fixturing element layout, clamping sequence and magnitudes in the clamping actuation sub-task could be accounted for implicitly in NC codes by a fixture planner before machining process starts. When a workpiece position probing system is employed to sense the displacement, the co-ordinate system can be shifted so as to enable the NC controller to correct the positional and orientational errors of the workpiece to some extent so that less dimensional errors of the finished workpiece would result.

In the clamping conservation sub-task, on the other hand, workpiece distortion within a fixture can be controlled by clamping actuation intensities during a machining operation. As the clamping intensities are carefully generated by a clamping optimisation algorithm during the planning stage, they are applied and adaptively varied by a dynamic clamp with respect to the instantaneous machining force magnitudes and directions. Hence, the in-process workpiece deformation is expected to be minimised.

6.2.2.2 System Overview

The core hardware of the intelligent fixturing system is a "live" fixture which would provide proactive reaction to the workholding process and attempt to correct the workpiece displacement and restrain the workpiece distortion as much as possible so as to guarantee the required accuracy for a workpiece. As illustrated in Figure 6.3, its two essential functions are tool path compensation and dynamic clamping actuation.

As all the ambient forces would cause the workpiece to displace within a fixture, it is best to perform tool path compensation both before and during the machining process. A variety of techniques such as inductive displacement sensing can be employed to carry out on-line measurement of workpiece displacement. However, an attempt to perform on-line tool path compensation requiring the presence of an open-architectured NC controller would be very difficult, if not impossible at this stage. Therefore tool path compensation is only performed before machining to correct the displacement incurred by clamping actuation, which will be elaborated on in Section 6.3 and would form a stepping stone to the in-process tool path compensation in future.

During machining, the intelligent fixture would execute such functions as fixture monitoring, dynamic clamping, regulatory control, and data acquisition so as to ensure the successful performance of machining operations and minimise workpiece deflection. Its schematic configuration is shown in Figure 6.4. The characteristic features of the intelligent fixture would include the following:

- Automated monitoring of the workpiece-fixture system

Automated monitoring refers to the identification of the characteristic changes of a workpiece-fixture system based on the evaluation of system signatures. It mainly consists of three tasks: sensing, signal processing and monitoring decision-making. Appropriate sensor signal sources include reaction forces on the fixturing elements, clamping forces, workpiece deflection, chattering and vibration. At this time, only reaction forces are used to monitor the stability of the workpiece-fixture system and as an input to the decision-making strategy for the self-configuration of the fixture system to provide the highest workpiece accuracy. The hardware required for performing this task includes force sensors and a data acquisition board.

- Applying dynamic clamping forces

Applying dynamic clamping forces refers to the action of adapting the clamping intensities according to the optimised fixturing parameters generated at the fixturing planning stage as well as the sensed information of workholding stability. The execution involves the co-ordination of the clamping action with the machining operation. Electromechanical or hydraulic actuators may be employed to apply the real-time optimal clamping forces to the workpiece, which will be discussed in Section 6.4.

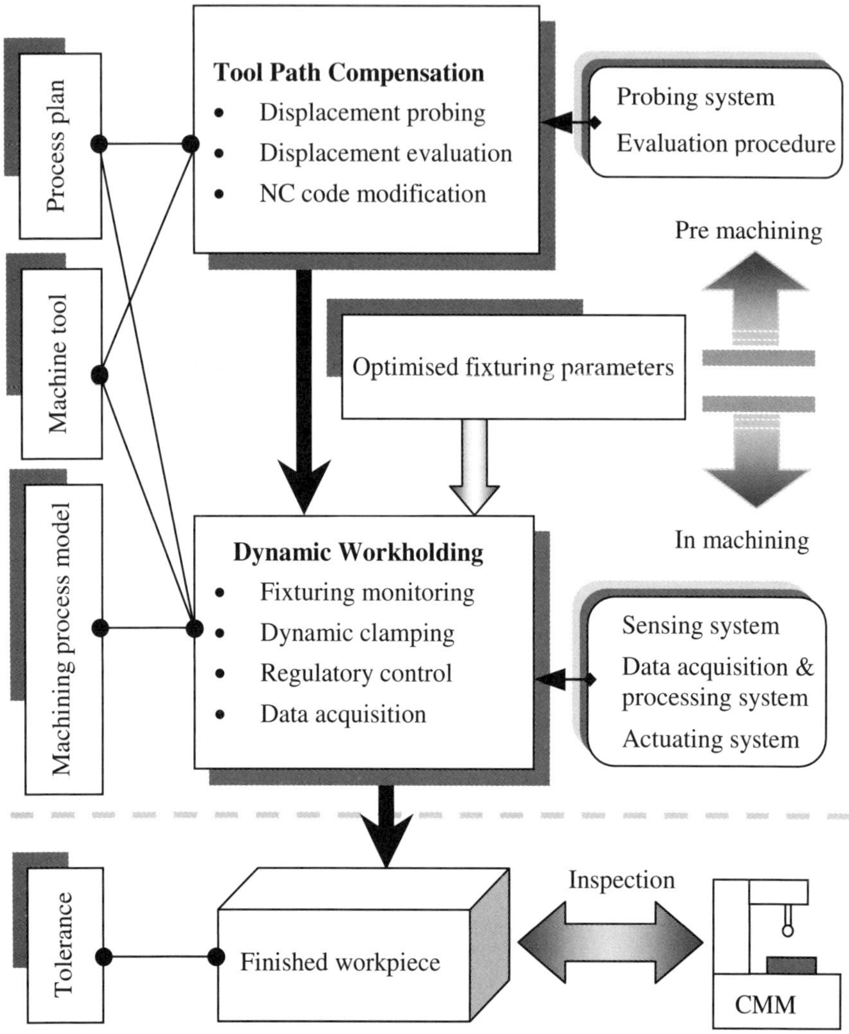

Figure 6.3 Functions of a "live" fixture for workholding execution

- Regulatory control of workholding stability

While automated monitoring tracks the execution of the workholding process and detects the departure from normal conditions, regulatory control applies preventive/corrective action with respect to clamping intensities in response to sudden undesirable changes in the workholding conditions due to external perturbations.

- Data acquisition

It refers to the collection and recording of fixturing process information and associated data validation, conditioning, smoothing, and corrective measures. A workholding history could be produced from the collected data, and significant workholding events (e.g., "lifting-off", "sliding", "vibration", "chatter", etc) and the time stamp of their occurrence can be identified through the evaluation of the sensed information. Consequently, corrections to the fixturing process or even the fixture design specifications can be made as necessary so as to fine tune the subsequent workholding operations.

The clamping features of the intelligent fixture are summarised in Table 6.1 with a comparison with other kinds of fixtures. In the ideal vision, a fully automated NC fixture would apply just enough sustainable locking forces to the workpiece in a self-adaptive manner. Such a visionary system would estimate the necessary clamping intensity in an on-line manner through sensory feedback and self-learning control strategy. It would also adapt the clamping positions as a fixed layout of clamping points may not be sufficient to support the workpiece from the point of view of workpiece deflection. An array of properly arranged clamping actuators may be employed to secure the workpiece, and each of them can be individually activated or deactivated by enabling signals sent from monitoring decision-making strategy such that the entire workpiece is well supported to prevent unfavourable deflection.

Table 6.1 Clamping features for various types of fixtures

Type of fixture	**Clamping planning**	**Clamping force optimisation**	**Clamping force magnitudes**	**Clamping positions**
Conventional fixture	no	No	constant	fixed
Advanced fixture	yes	No	constant	fixed
Intelligent fixture	yes	Yes	dynamic	fixed
Fully automated CNC fixture (visionary)	yes	Yes	dynamic	dynamic

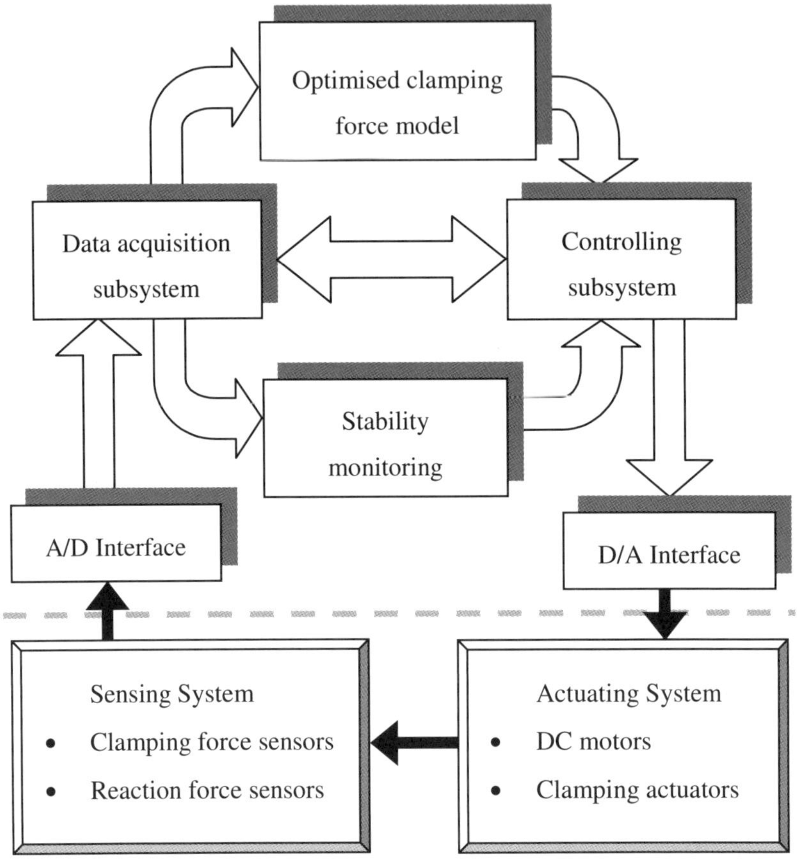

Figure 6.4 Schematic layout of a "live" fixture configuration

6.3 Tool Path Compensation

6.3.1 Determination of Workpiece Displacement

In order to perform the tool path compensation procedure to fit the actual workpiece position, the deviation of workpiece co-ordinate system in both translational and rotational terms with respect to the machine co-ordinate system must be determined.

It is reasonable to assume that the workpiece is located by a fixture in the machine workspace exactly as planned, meaning that the workpiece co-ordinate system coincides with the machine co-ordinate system before applying clamping forces. After clamping actuation, the workpiece undergoes small displacement due to the fixture flexibility, the deformation of the small area contacts between the workpiece and fixturing elements and/or lift-off of the workpiece. Probes such as digital dial gauges can be used to measure the displacement phenomenon (Figure 6.5).

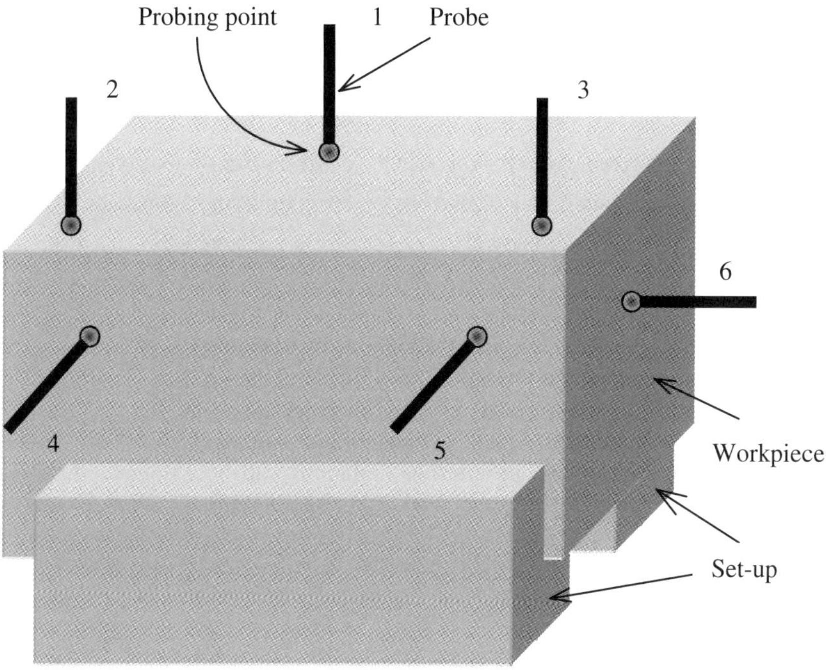

Figure 6.5 Probing of workpiece displacement

One of the practical probing schemes is to probe six points on the three surfaces directly opposite to the locating data: three probing points on the top surface, two points on the side surface and one point on the end surface. Such a probing arrangement ensures that the sensors would produce six linearly independent readings. Probing points should preferably be placed on clear cut surfaces produced by a previous operation, and the further apart these points, the better results one can obtain.

Supposing the position vector and orientation vector of a feature on the workpiece in the machine co-ordinate system are $\boldsymbol{r}_m = [\boldsymbol{x}_m \quad y_m \quad z_m]^T$ and

$\theta_m = [\theta_{xm} \quad \theta_{ym} \quad \theta_{zm}]^T$ respectively, and in the workpiece co-ordinate system are $r = [x \quad y \quad z]^T$ and $\theta = [\theta_x \quad \theta_y \quad \theta_z]^T$ respectively, the relationship between them can be described by a transformation matrix when the workpiece undergoes a rigid body displacement after clamping actuation:

$$\begin{aligned} \boldsymbol{r}_m &= \boldsymbol{T}\,\boldsymbol{r} + \Delta\boldsymbol{r} \\ \theta_m &= \boldsymbol{T}\,\theta \end{aligned} \tag{6.1}$$

where $\boldsymbol{T}$ is a 3 by 3 rotation matrix and $\Delta\boldsymbol{r}$ is a translation vector $\Delta\boldsymbol{r} = [\Delta x \quad \Delta y \quad \Delta z]^T$.

$$\boldsymbol{T} = \begin{bmatrix} \cos\beta\cos\gamma & \sin\alpha\sin\beta\cos\gamma - \cos\alpha\sin\gamma & \cos\alpha\sin\beta\cos\gamma + \sin\alpha\sin\gamma \\ \cos\beta\sin\gamma & \sin\alpha\sin\beta\sin\gamma + \cos\alpha\cos\gamma & \cos\alpha\sin\beta\sin\gamma - \sin\alpha\cos\gamma \\ -\sin\beta & \sin\alpha\cos\beta & \cos\alpha\cos\beta \end{bmatrix} \tag{6.2}$$

where

Δx = translation along the x-axis; α = rotation about the x-axis,
Δy = translation along the y-axis; β = rotation about the y-axis,
Δz = translation along the z-axis; γ = rotation about the z-axis.

As the displacement is small in magnitude compared with the overall dimensions of the workpiece, infinitesimal rotations can be assumed, therefore, higher-order terms may be ignored, and the transformation matrix $\boldsymbol{T}$ can be approximated as follows:

$$\boldsymbol{T}' = \begin{bmatrix} 1 & -\gamma & \beta \\ \gamma & 1 & -\alpha \\ -\beta & \alpha & 1 \end{bmatrix} \tag{6.3}$$

Consequently the position deviation of a point on the workpiece due to the workpiece displacement is as follows:

$$\delta\boldsymbol{r} = \boldsymbol{r}_m - \boldsymbol{r} = (\boldsymbol{T}' - \boldsymbol{I})\,\boldsymbol{r} + \Delta\boldsymbol{r}$$

$$\begin{pmatrix} \delta x \\ \delta y \\ \delta z \end{pmatrix} = \begin{bmatrix} 0 & -\gamma & \beta \\ \gamma & 0 & -\alpha \\ -\beta & \alpha & 0 \end{bmatrix} \begin{pmatrix} x \\ y \\ z \end{pmatrix} + \begin{pmatrix} \Delta x \\ \Delta y \\ \Delta z \end{pmatrix} \tag{6.4}$$

or

$$\begin{aligned}\delta x &= \Delta x - y\gamma + z\beta \\ \delta y &= \Delta y + x\gamma - z\alpha \\ \delta z &= \Delta z - x\beta + y\alpha\end{aligned} \tag{6.5}$$

Substituting the measured deviation values at the six probing points into Equation 6.5, the displacement of the workpiece can be determined by the following equation:

$$\begin{pmatrix}\Delta x\\ \Delta y\\ \Delta z\\ \hdashline \alpha\\ \beta\\ \gamma\end{pmatrix} = \begin{bmatrix} 0 & 0 & 1 & y_1 & -x_1 & 0 \\ 0 & 0 & 1 & y_2 & -x_2 & 0 \\ 0 & 0 & 1 & y_3 & -x_3 & 0 \\ 0 & 1 & 0 & -z_4 & 0 & x_4 \\ 0 & 1 & 0 & -z_5 & 0 & x_5 \\ 1 & 0 & 0 & 0 & z_6 & -y_6 \end{bmatrix}^{-1} \begin{pmatrix}\delta z_1\\ \delta z_2\\ \delta z_3\\ \delta y_4\\ \delta y_5\\ \delta x_6\end{pmatrix} \tag{6.6}$$

where $(x_i \quad y_i \quad z_i)$ are the co-ordinates of the probing points and δx, δy, δz are the linear deviations along the corresponding axes at the probing points.

6.3.2 Workpiece Displacement by Experimental Measurement

Experimental investigation of the effect of clamping actuation on displacement of workpiece within a fixture has been carried out. The testing set-up is similar to the one presented in the Figure 5.8 of Chapter 5, where a solid block of 220 by 130 by 80 mm is located by six flat-ended cylindrical locator pins (∅20mm) and secured by two motorised side clamps. The dimensions ensure that the workpiece is structurally rigid. The secondary and tertiary locators are mounted onto angle plates which are fastened to the fixture baseplate while the primary locators are directly connected to the threaded holes on the baseplate.

Clamping forces are measured with Kistler piezoelectric force sensors. Each clamp applies 600 N of clamping force sidewise onto the workpiece during testing, and the displacements on the probing points are measured with dial test indicators having accuracy of 0.002 mm per division. Two fixturing configurations and three clamping sequence schemes (one concurrent and two sequential) are investigated.

The only difference between the first testing case (Figure 6.6) and the second testing case (Figure 6.7) is the arrangement of primary locators. The layout of the measuring points is identical in both cases. Three clamping sequence schemes have been tested: (a) both clamps C1 and C2 are applied simultaneously, (b) clamp C1 is actuated first, then clamp C2, and (c) clamp C2 is actuated first, then clamp C1. The measured deviation values and the evaluated displacements for each case are tabulated in Tables 6.2, 6.3 and Tables 6.4, 6.5 respectively.

The following observations can be drawn from the testing results:

- Workpiece displacement during clamping actuation is non-trivial and cannot be ignored in the machining error analysis. (Note that the largest measured displacement component is 0.189mm!) To reduce the magnitude of workpiece displacement, a prudent tool designer should appropriately choose: (1) contact surface geometry of the fixturing elements, (2) the number and type of supports and their placement, (3) fixturing element placement, (4) clamping intensities, (5) clamping sequence, and (6) rigid frame and structure for a fixture, etc.

- Angular deviations (α, β, γ) are much smaller in general than linear deviations (Δx, Δy, Δz). However, insignificant angular deviations may cause significant changes in the positions of individual features when they are positioned far away from the co-ordinate origin.

- Out of the angular deviations, the γ component about z-axis is always the smallest (about one order smaller). This is because the workpiece is rigidly constrained in the xy-plane by the side clamps in the fixturing configuration.

- The linear deviation component Δz in the first case is about one order smaller than that in the second case. It is shown that Δz is, to a great extent, affected by the placement of the primary locators, whereas the other linear components Δx, Δy are less sensitive to it.

- The linear deviation component Δx is usually larger than the linear error Δy. This can be explained by the fact that the fixture rigidity in y-axis direction is stronger than that in x-axis direction (two locators versus one locator), and hence less y-direction deflection would be induced by the clamping force.

- For both testing cases, the largest linear displacement components (0.189 mm and 0.185 mm) occur when two clamps are applied concurrently. This interesting result suggests that a concurrent clamping scheme may not always be desirable and favourable with respect to the minimisation of the workpiece deviations, even when simultaneous clamping is possible. Within the context of this experiment, it is also inconclusive to suggest a preferred clamping sequence that would minimise both the linear and angular deviations.

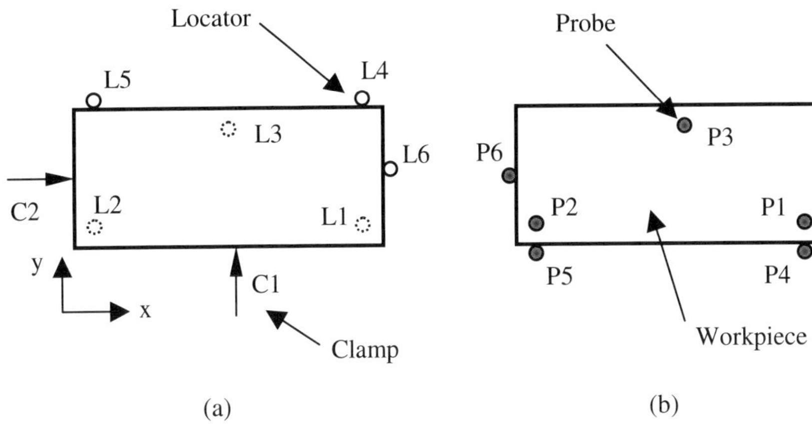

Figure 6.6 (a) Layout of locators and clamps and (b) layout of measurement points (case A)

Table 6.2 Measurement of position deviations at the probing points (averages of five readings) (case A)

Test No.	Clamping scheme	Measured deviations (mm)					
		P1	P2	P3	P4	P5	P6
		δz_1	δz_2	δz_3	δy_4	δy_5	δx_6
1	Concurrent	-0.023	-0.047	+0.182	+0.041	+0.053	+0.095
2	C1 → C2	+0.108	-0.003	+0.042	+0.036	+0.041	+0.094
3	C2 → C1	-0.002	-0.004	+0.089	+0.049	+0.037	+0.058

Table 6.3 Evaluation of workpiece displacement after clamping actuation (case A)

Test No.	Evaluated displacement					
	Translational (mm)			Rotational (degree)		
	Δx	Δy	Δz	α	β	γ
1	+0.099	+0.189	-0.056	+0.103	-0.006	-0.003
2	+0.132	+0.034	-0.005	-0.005	-0.030	-0.001
3	+0.062	+0.094	-0.007	+0.043	-0.001	+0.003

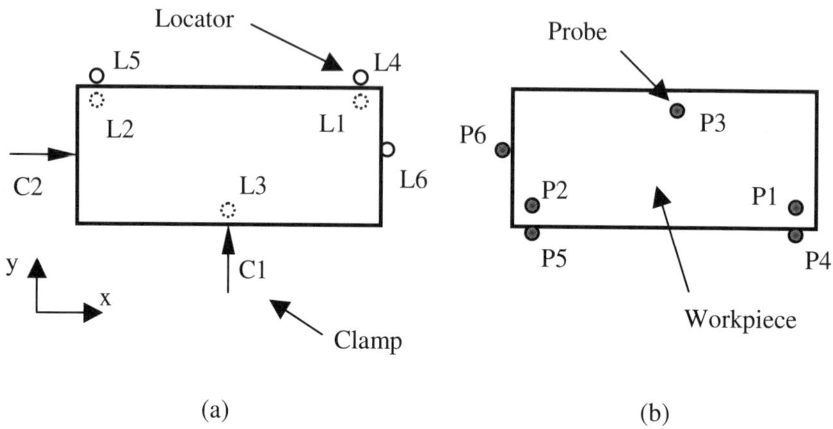

Figure 6.7 (a) Layout of locators and clamps and (b) layout of measurement points (case B)

Table 6.4 Measurement of position deviations at the probing points (averages of five readings) (case B)

Test No.	Clamping scheme	Measured deviations (mm)					
		P1	P2	P3	P4	P5	P6
		δz_1	δz_2	δz_3	δy_4	δy_5	δx_6
1	Concurrent	+0.099	-0.119	+0.114	+0.049	+0.028	+0.101
2	C1 → C2	+0.105	-0.114	+0.067	+0.064	+0.039	+0.058
3	C2 → C1	+0.071	-0.082	-0.108	+0.042	+0.030	+0.067

Table 6.5 Evaluation of workpiece displacement after clamping actuation (case B)

Test No.	Evaluated displacement					
	Translational (mm)			Rotational (degree)		
	Δx	Δy	Δz	α	β	γ
1	+0.185	+0.105	-0.129	+0.059	-0.059	+0.005
2	+0.143	+0.083	-0.122	+0.034	-0.059	+0.006
3	+0.125	-0.034	-0.081	-0.048	-0.041	+0.003

6.3.3 Tool Path Compensation

The task of tool path compensation is essentially dependent on the capability of a machine tool. It is constrained by the hardware limitations of the machine tool. On a 3-axis NC machine tool, the linear deviations of a feature can be offset through tool path adjustment and the angular deviations can only be partially compensated, whereas both linear and angular deviations can be fully corrected on a 5-axis NC machine tool.

6.3.3.1 Linear Compensation

The effect of workpiece displacement on machining errors as well as the benefits of tool path compensation can be illustrated in a milling operation. Supposing a 3-axis NC machine is used to produce a through slot on a block, which is located by six locators and held by two side clamps. After loading onto the fixture, the block workpiece makes contact with the locators and hence establishes a deterministic position with respect to the machine co-ordinate system, which, in this case, is coincident with the workpiece co-ordinate system as shown in Figure 6.8.

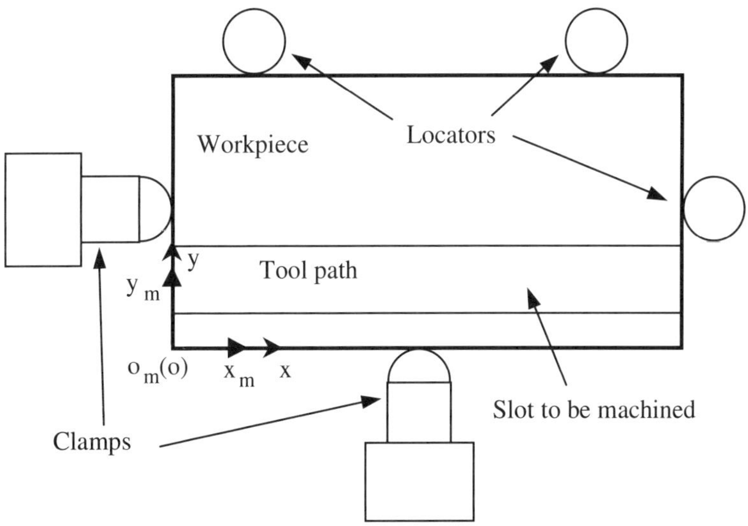

Figure 6.8 Workpiece establishes a deterministic position by locators before clamping actuation

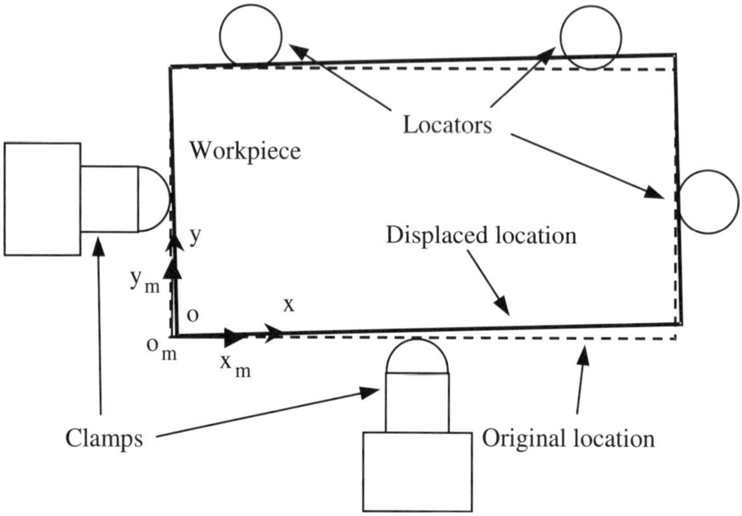

Figure 6.9 Workpiece displacements within a fixture after clamping actuation

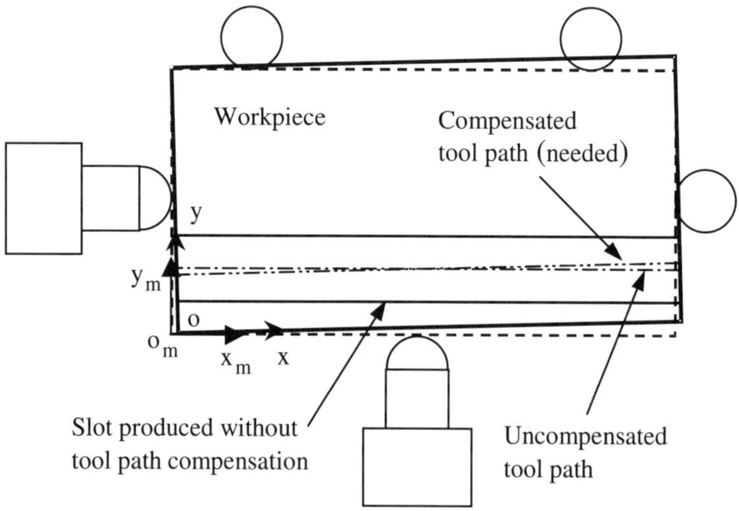

Figure 6.10 Machining error of the finished feature caused by workpiece displacement

After the clamping forces are activated, the workpiece is displaced within the fixture and the workpiece co-ordinate system deviates from the machine co-ordinate system (Figure 6.9). If the cutting path remains uncompensated, it is parallel to the machine reference frame, but out of line with the desired tool path on the workpiece, resulting in the machining error of the finished surface (Figure 6.10). In order to overcome the divergence, tool path has to be compensated through a transformation same as the one experienced by the workpiece.

Supposing the workpiece displacement has been evaluated through a probing procedure and characterised by a rotation matrix $\boldsymbol{T}^e$ and a translation vector $\Delta\boldsymbol{r}^e$, and the starting and end points of the linear tool path are represented by $\boldsymbol{r}^1$ and $\boldsymbol{r}^2$ respectively, the new starting and end points $\boldsymbol{r}_m^1$ and $\boldsymbol{r}_m^2$ after compensation are computed as follows:

$$\begin{aligned} \boldsymbol{r}_m^1 &= \boldsymbol{T}^e\,\boldsymbol{r}^1 + \Delta\boldsymbol{r}^e \\ \boldsymbol{r}_m^2 &= \boldsymbol{T}^e\,\boldsymbol{r}^2 + \Delta\boldsymbol{r}^e \end{aligned} \tag{6.7}$$

As shown in Figure 6.10, the compensated tool path is no longer parallel to the machine reference frame. NC programme codes can be easily modified to enable the machine tool to perform the updated tool path.

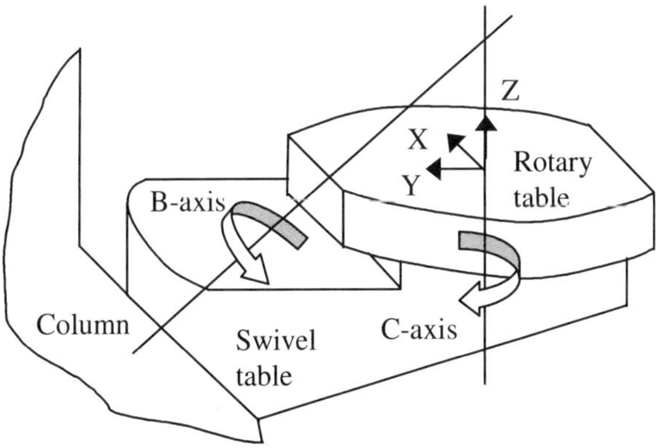

Figure 6.11 Movement axes and machine co-ordinate system in a 5-axis machine tool

6.3.3.2 Angular Compensation

The complete compensation of the workpiece displacement can only be performed on a 5-axis machine tool. In the experimental work, a Deckel Maho DMU 50V 5-

axis universal milling machine, which is fitted with a swivel table and a rotary table to provide two additional rotational axes (B-axis and C-axis), is used (Figure 6.11). Supposing a hole feature is to be drilled by the machine tool, and the starting point of the hole is the position $\boldsymbol{r}$ of the feature and the axis of the hole is the orientation of the hole (Figure 6.12), the new position $\boldsymbol{r}_m$ and orientation θ_m of the hole feature relative to the machine reference frame is expressed as follows when the workpiece undergoes a small displacement:

$$\begin{aligned} \boldsymbol{r}_m &= \boldsymbol{T}^e\, \boldsymbol{r} + \Delta \boldsymbol{r}^e \\ \theta_m &= \boldsymbol{T}^e\, \theta \end{aligned} \tag{6.8}$$

where $\boldsymbol{T}^e$ and $\Delta \boldsymbol{r}^e$ are the angular matrix and translation vector based on the evaluated displacement of the workpiece, respectively.

It is shown in Figure 6.12 that, if the machine table remains unchanged in its orientation, the tool axis would no longer be concentric with the nominal hole axis, resulting in the machining error of the hole feature in the form of diameter error, perpendicularity error and parallelism error (Figure 6.13). This eccentric angle cannot be offset by a linear compensation, it can only be compensated by a careful rotation of the machine table. In the Deckel Maho machine, it can be achieved through a combined rotation movement of both the rotary table and swivel table after the deviated orientation θ_m of the hole feature is converted into the individual rotation angles of B axis (swivel table) and C axis (rotary table). The conversion data can be computed by AxTra software (Deckel, 1996), and a sample NC code section for performing the compensated tool path movement looks like below:

```
. . .                                        Swivelling angle   Rotary angle
N50 G17      (Main Plane XY, Tool Z)
N51 G51      (Cancel Set Zero Point)
N52 G54      (Activate Zero Point Shift)
N53 G93 X100.202 Y-64.910 Z75.181 B0.146 C0.052 B4=0.052
N54 T1 M10                        (Change Drill Tool)
N55 G81 Z-20 F1000 S2500 M13      (Define Drill Cycle)
N56 G79 X0 Y0 Z0 B0 C0            (Activate Drill Cycle)
N57 G0 Z250                       (Rapid Tool Return)
. . .
```

After the compensating rotation movement, the tool axis becomes concentric with the required hole axis again (Figure 6.12) and therefore the angular deviations are completely offset by modifying the NC programmes.

In general, this compensating procedure can be repeated for multiple features on a displaced workpiece. However, it should be noted that a collision-free tool path might not be safe any more after the rotating movement of the machine table. Tool interference must be re-checked for the tool path compensation.

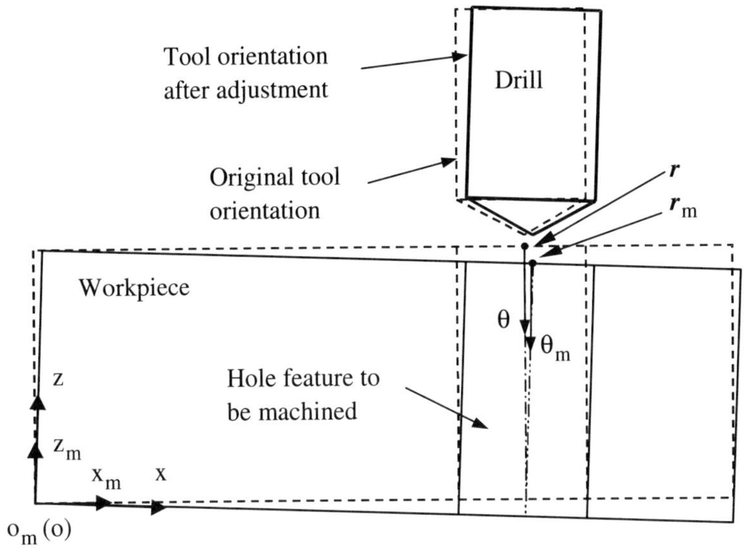

Figure 6.12 Compensated and uncompensated tool orientation

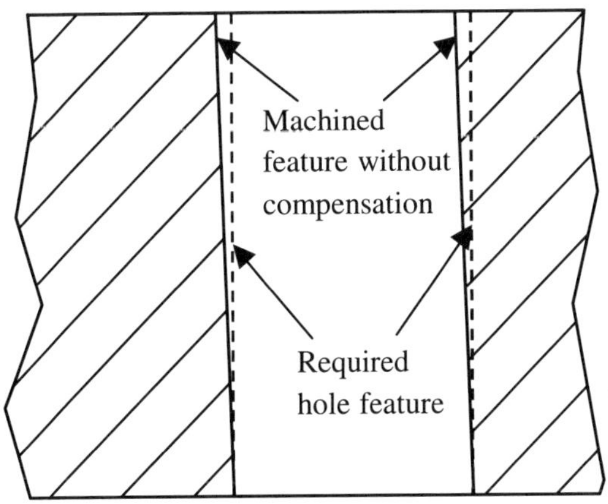

Figure 6.13 Machining error of a hole feature caused by workpiece displacement

6.4 Dynamic Clamping Actuator

6.4.1 Fixturing Error Control Strategy

To control fixturing errors in terms of workpiece displacement and deformation, a fixture planner has the following options to choose from: the number of fixturing elements to contact the workpiece, the placement of these elements, the type and contact geometry of these elements, and the clamp actuation intensities. The first three variables may be considered as "firmware" approaches while the last one as "software" approach.

The objective here is to minimise displacement and deformation so as to improve workpiece quality. Intuitively, displacement and distortion could be reduced with the application of a large number of fixturing elements. This would ensure that clamping and machining loads are distributed over many contact regions, reducing both contact region deformation and workpiece deflection. However, this approach is usually not practical. The presence of redundant locators leads to greater positioning variability during set-up. This variability, in turn, may cause greater error than workpiece displacement. Likewise, the utility of a large number of clamps and supports ultimately increases the probability of tool-path interference and set-up time. Oversized contact geometry design would also give rise to tool path clearance problem. In contrast, careful specification and prudent actuation of the last parameter of clamping intensities would have significant impact on workpiece displacement and distortion, cycle time, and surface finish. Therefore, a "software" approach to the containment of fixturing errors is preferred over the "firmware" methods.

6.4.2 Need for Dynamic Clamping Force

Clamping force is actively applied to a workpiece after it is loaded onto a fixture. If a workpiece is totally restrained by form closure without counting on friction, a substantial clamping force is not needed. However the friction force, in practice, is usually relied upon to resist the cutting force components. Therefore, a substantial clamping force is needed to hold a workpiece with the presence of friction.

In a conventional fixture, clamping force is kept constant in both magnitude and direction simply for the ease of operation, implying that the clamping magnitude is prepared for the worst machining condition. The problem is that it is very hard to predict when and where the worst case would occur. Therefore, constant clamping intensity has to be excessively large for the whole operation, just to be on the safe side. That means over-clamping is always exercised where otherwise only a lower clamping intensity is needed. In contrast, dynamic clamping intensity entails that the clamping force is varying with time rather than fixed all the time, thus provides minimal sustainable clamping forces at every moment and successively reduces the in-process distortion of the workpiece.

6.4.3 Dynamic Clamping Actuator

The schematic of a developed dynamic clamp prototype is shown in Figure 6.14 where the electromechanical system primarily consists of a DC servomotor and a linear actuator. The motor is the driving power source and its position and speed are manipulated by a digital controller. The motor provides high torque to the actuator which, in turn, transforms the rotational motion into a linear movement. The actuator operates with a high reduction worm gear that has the ability to self-lock and maintain its position and the force. A force sensor is embedded inside the end-effector of the actuator to feed back the force value imparted to the workpiece. An encoder is attached to the rear end of the motor shaft to feed back the rotational position and speed, forming a local closed-loop for maintaining high accuracy of motion control.

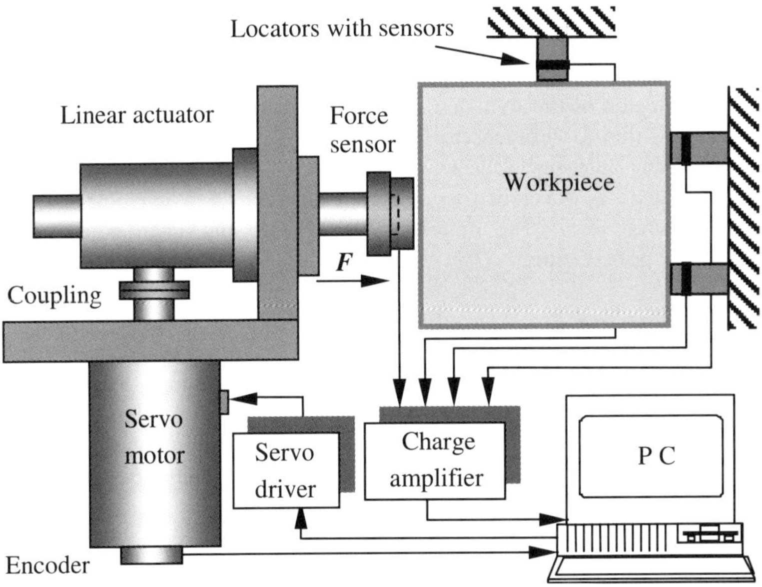

Figure 6.14 Computer control schematic for dynamic clamping actuator

6.4.3.1 Sensor Integration

Sensor integration is critical to the development of a dynamic actuator. The presence of sensors allows real-time information transfer between the fixturing action and its controller. The purpose of sensor integration into both locating and clamping components is three-fold: indication of the registration of a workpiece within the fixture before clamping, monitoring of clamping actuator such as detection of clamp malfunction during machining, closed-loop control of the dynamic clamping intensity. Piezoelectric sensors are employed in the measurement of dynamic forces for their high resolution, rigidity, linearity and compactness.

6.4.3.2 Digital Interfacing

The task of sensing and controlling clamping operation is performed using a Pentium personal computer. In this scheme, analogue signals from the force sensors are received and conditioned through an analogue to digital converter (ADC) located inside the PC, and the digital command signals are sent from the PC to the clamp for the initiation, on-line adjustment, and termination of the clamping operation through another digital to analogue converter (DAC). The input/output signal processing and conditioning and the control of the clamping process are carried out with appropriate control software.

6.4.3.3 Clamping Force Control

Since clamping force is affected by the contact condition between the workpiece and the end-effector of the dynamic actuator, it appears reasonable to adjust the end-effector position to compensate for any changes in the force value to ensure that the clamping intensity follows a reference profile closely. An appropriate controlling strategy is developed to deliver a desirable dynamic response and to regulate the force level. The details of the controlling methodology will be discussed in the next section.

6.4.3.4 Clamping Task Programmability

In order to accommodate the multiple machining operations performed on a workpiece within one set-up, it is important that the process of clamping actuation is programmable in terms of clamping sequence, actuation duration, clamping intensity spectrum. In this way, all the functions of the clamping process can be tailored to fulfil the requirements for securing a workpiece.

6.5 Clamping Force Control Strategy

Force control with the position of the clamping end-effector is achieved using a servo drive system. The implemented controller and its corresponding block diagram representation of the control process are illustrated in Figures 6.15 and 6.16 respectively. The force feedback serves as a position correction and is subtracted from the force reference provided by the clamping force optimisation model. The resulting force error is then converted into a relative position through the multiplication of the reciprocal of compliance, $\boldsymbol{k_e}$, of the controlled system. Since $\boldsymbol{k_e} = \boldsymbol{dF/dx}$, the compliance can be estimated as follows:

$$k_e(n) = \frac{y(n) - y(n-1)}{x(n-d) - x(n-d-1)} \qquad (6.9)$$

where $\boldsymbol{y(n), x(n)}$ are values of the measured clamping force and the position of end-effector respectively at time $\boldsymbol{t = nT}$ ($\boldsymbol{T}$ is the sampling interval), and $\boldsymbol{d}$ is the servo system time delay in number of samples. This method attempts to combine simplicity and robustness of the impedance method with the ability of simultaneously handling both position and force variables typical of a hybrid control approach.

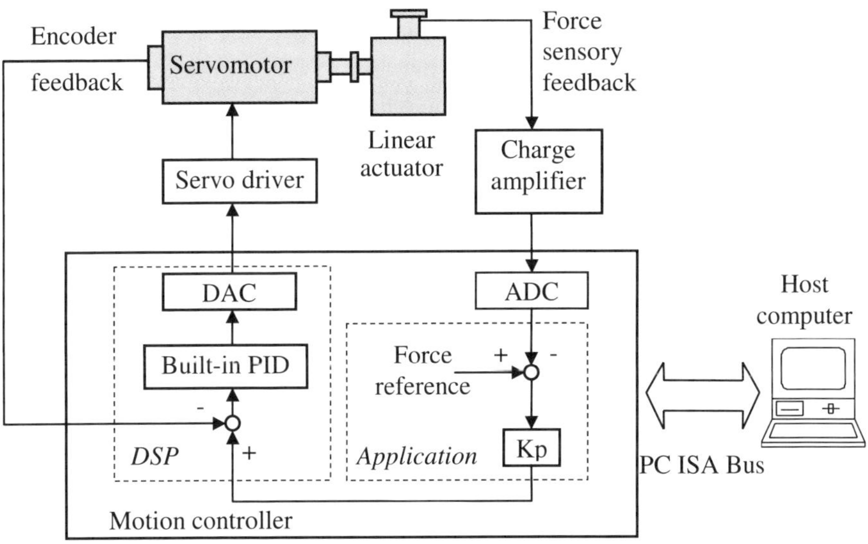

Figure 6.15 Implemented controller for dynamic clamping force

The input to the system is voltage $\boldsymbol{u(n)}$ and the output is the force value $\boldsymbol{y(n)}$ at the end-effector (Figure 6.16). The position and speed of the servomotor are treated as intermediate variables so that the process can be modelled as a single-input-single-output (SISO) system. The clamping force imparted to the workpiece is controlled by manipulating the position of the actuator. Clamping force is measured by a piezoelectric sensor that was previously calibrated for its linear relationship between applied force and generated voltage. The DC servomotor is linked to a local position and speed control loop. The Pentium PC acts as the master controller and is interfaced to the servo controller via a PC bus link.

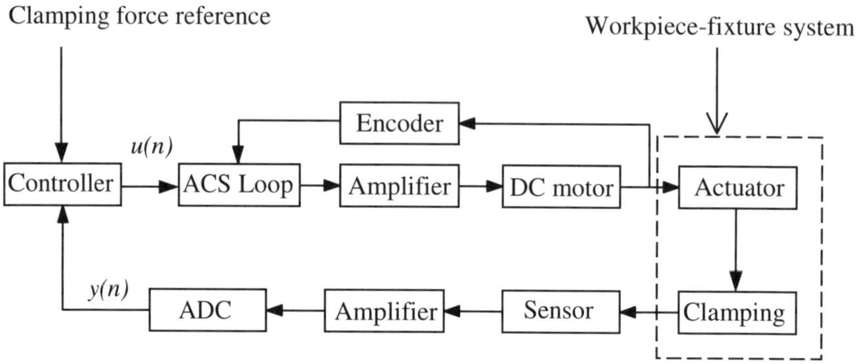

Figure 6.16 A block diagram representation of the controlling system

6.5.1 DC Motor Servo Controller

As shown in the block diagram representation of the overall control system in Figure 6.16, the control architecture consists of two feedback loops, viz., local position feedback from the encoder and global force feedback from the piezoelectric sensor. The local closed-loop is manipulated using a general-purpose servo controller (ACS SB214). It has a built-in PID action controller, or digital filter. The purpose of the digital filter is to modify the open loop transfer function of the servo system in order to improve the time response, stability and stiffness of the servo loop. A second order standard filter is employed in the following form:

$$\boldsymbol{D}(z) = \boldsymbol{G} \cdot \frac{z - z_1}{z + \boldsymbol{p}_1} \cdot \frac{z - z_2}{z - 1} \tag{6.10}$$

where $\boldsymbol{G}$ is the total gain, the second part is a lead-lag term for the stability of the servo loop, and the third part is an integration term which increases the stiffness of the system, improves the tracking accuracy and reduces the steady state error to zero. The parameters (zeros z_1, z_2 and pole $\boldsymbol{p}_1$) of the digital filter are appropriately designed. Their values can be fine-tuned after pilot running through a tuning session. The hierarchical structure with two feedback loops gives control flexibility, provides possibilities of fast and yet robust control, and ensures the required force is applied.

6.5.2 System Identification of Process

The fundamental requirement for a control system is to characterise the dynamic behaviour of a process to be controlled. A realistic parametric model is needed and consequently the first step is system identification. The stochastic modelling technique can be employed to identify the system dynamics, in which known inputs are applied to the system and the outputs are observed for subsequent model development. The system can be assumed to have the form:

$$\begin{aligned} y(n) = -a_1 y(n-1) - a_2 y(n-2) - \cdots - a_k y(n-k) + \\ + b_0 u(n-d) + b_1 u(n-1-d) + \cdots + b_k u(n-k-d) + \varepsilon(n) \end{aligned} \quad (6.11)$$

where $y(n)$, $u(n)$ are values of the measured clamping force and the manipulated voltage respectively at time $t = nT$; a_i and b_i are model coefficients; k is the order of the process model; d is any time delay in number of samples; $\varepsilon(n)$ is the value of any noise corrupting the measurements at time t, this term includes small unpredictable changes in the process and errors in the output due to modelling and is assumed to be statistically independent of the input signal $u(n)$ and generated by passing a white noise $e(n)$ through a stable linear filter.

The difference equation (6.11) can be rewritten as follows:

$$y(n) = \frac{B}{1+A} u(n) + \frac{C}{1+A} e(n) \quad (6.12)$$

where

A = $a_1 z^{-1} + a_2 z^{-2} + \cdots + a_k z^{-k}$

B = $b_0 z^{-d} + b_1 z^{-1-d} + \cdots + b_k z^{-k-d}$

e_n = a white Gaussian noise sequence with zero mean

z^{-i} = the backward shift operator defined by $z^{-i} u(n) = u(n-i)$

C = a polynomial with the same form as A and B and filters the white noise $e(n)$ to describe the measured random signal noise $\varepsilon(n)$.

For simplicity, $C = 1$ is assumed here and is equivalent to the assumption of an integral of white noise to the system. The above model for a random process disturbance is termed an Auto Regressive Moving Average (ARMA) model.

Determination of the system parameters is an essential stage in modelling. They can be estimated using the least squares technique. In order to provide input-output data for modelling, the manipulated signal is varied in an arbitrary manner between limits to the control programme. Both the signal and the sampled measurements of the resulting clamping force are observed and stored on hard disk.

Using mathematical tools such as MATLAB (Kuo, 1994), the resulting discrete time model of the process is obtained in the form as follows:

$$y(n)=\frac{(b_1z^{-1}+b_2z^{-2}+b_3z^{-3})u(n)+e(n)}{1+a_1z^{-1}+a_2z^{-2}+a_3z^{-3}} \tag{6.13}$$

6.5.3 Generalised Minimum Variance Model

The aim of a generalised minimum variance (GMV) self-tuning controller is to minimise the variance of the output of a system when subjected to a stochastic disturbance. Its schematic structure is shown in Figure 6.17 where a difference ***v(n)*** between actual output **y(n)** and the desired output ***R(n)*** is introduced to be the pseudo-output of the system. Consequently the control objective is shifted to the task of minimising the variance of the pseudo-output ***v(n)***.

The parameters ***A*** and ***B*** can be estimated using the recursive least squares algorithm from

$$v(n)=-\hat{a}_1v(n-1)-\hat{a}_2v(n-2)-\hat{a}_3v(n-3)+\hat{b}_1u(n-1)+\hat{b}_2u(n-2)+\hat{b}_3u(n-3) \tag{6.14}$$

where the presence of a circumflex (^) indicates an estimation. Therefore, the next output can be predicted on the basis of $\hat{A}$ and $\hat{B}$ from

$$\hat{v}(n+1)=-\hat{a}_1v(n)-\hat{a}_2v(n-1)-\hat{a}_3v(n-2)+\hat{b}_1u(n)+\hat{b}_2u(n-1)+\hat{b}_3u(n-2) \tag{6.15}$$

Setting the predicted pseudo-output of $\hat{v}(n+1)$ to zero leads to the GMV control algorithm:

$$u(n)=\frac{1}{\hat{b}_1}\left\{\hat{a}_1v(n)+\hat{a}_2v(n-1)+\hat{a}_3v(n-2)-\hat{b}_2u(n-1)-\hat{b}_3u(n-2)\right\} \tag{6.16}$$

If the value of $\hat{b}_1$ is very small, then numerical problems will arise and the ***u(n)*** from Equation 6.16 may be very large. This problem is solved by providing a fixed parameter g_1 to replace $\hat{b}_1$. The estimation model is now given by:

$$v(n)-g_1u(n-1)=-\hat{a}_1v(n-1)-\hat{a}_2v(n-2)-\hat{a}_3v(n-3)+\hat{b}_2u(n-2)+\hat{b}_3u(n-3) \tag{6.17}$$

And substituting g_1 into Equation 6.17 yields the GMV control law:

$$u(n) = \frac{1}{\hat{g}_1}\left\{\hat{a}_1 v(n) + \hat{a}_2 v(n-1) + \hat{a}_3 v(n-2) - \hat{b}_2 u(n-1) - \hat{b}_3 u(n-2)\right\} \qquad (6.18)$$

Consequently, the simplified GMV control strategy can be summarised as follows:

1. Provide a pre-defined value of g_1 in Equations 6.17 and 6.18 based on experimental testing or empirical recommendation.

2. At each sampling interval, estimate the model parameters of $\hat{a}_1$, $\hat{a}_2$, $\hat{a}_3$, and $\hat{b}_2, \hat{b}_3$ using Equation 6.17 in recursive fashion.

3. Measure the clamping force $y(n)$ and then compute the pseudo-output $v(n)$.

4. Determine the manipulated variable $u(n)$ from Equation 6.18.

5. Go back to Step 2 and repeat.

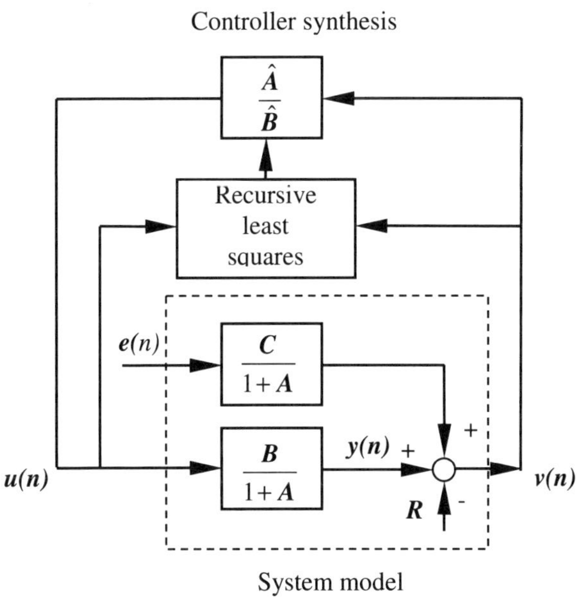

Figure 6.17 Generalised minimum variance self-tuning control scheme

6.6 Experimental Facility

This section presents the experimental set-up, testing procedure, and observational results with discussions. The experimentations are intended to validate the clamping force control algorithm and to determine the impact of the intelligent fixturing system on the workpiece quality and machining process. The test facility for performing these experiments is first introduced. Next, validation testing is presented in which the proposed force control strategy is established. Finally, comparative experiments are carried out to demonstrate the advantages of a dynamic fixturing execution. Measured geometric errors are compared with and without using the dynamic clamping forces.

The experimental facility employed to carry out the tests consists of several components. This includes an NC machining centre, the developed "live" fixture, workpieces, dial indicators, charge amplifiers, and a personal computer with a data acquisition system and a force control system. Both the acquisition and control system refers to the hardware and related software. In addition to those components, a co-ordinate measuring machine (CMM) is used to inspect the workpiece accuracy after the machining operation.

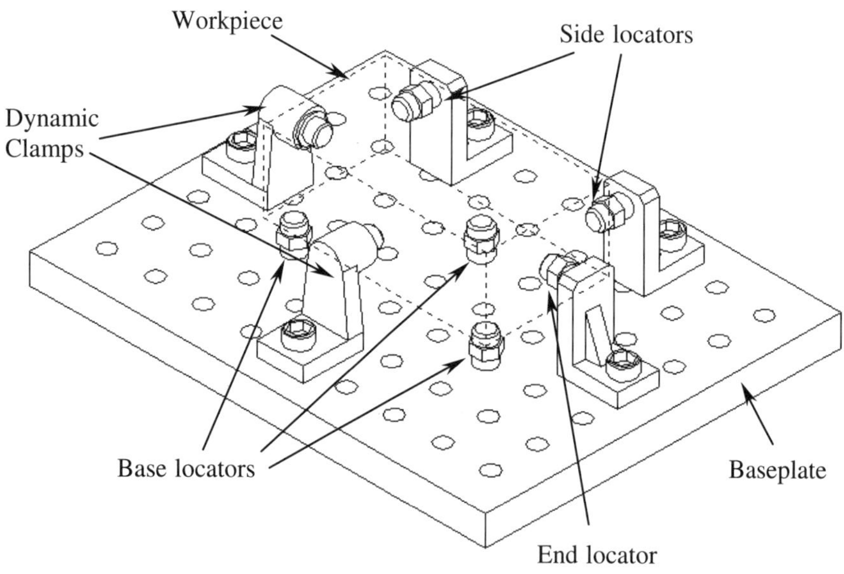

Figure 6.18 Schematic layout of the test fixture in isometric view

The machine tool used in the experiments is a Deckel Maho model DMU 50V 5-axis universal milling machining centre. This machining centre is selected for the reason that both linear and angular tool path compensation can be accomplished. The resolution of the axis movement is 1μm and the measured linear accuracy by a laser interferometer is ±3.5μm.

The developed "live" fixture is built upon a baseplate with a matrix of precisely positioned holes such that the placement of both locators and clamps can be easily reconfigured for different workpiece dimensions and shapes. An isometric view of the fixture configuration is shown in Figure 6.18. There are six locating pins and two dynamic clamps, each of them is embedded with a Kistler force sensor and its associated charge amplifier is also a Kistler 5038A3 model. Each clamp is electromechanically driven and can be independently controlled with a maximum capacity of 2500 Newtons. Actuated clamping intensity could be continuously changed.

A Data Translation DT31-EZ multi-function board is used for the task of data acquisition. The software that operates the data acquisition is HP-VEE version 3.1 (Helsel, 1994). An ACS SB214 multi-axis motion controller board is employed for controlling the clamping force actuation.

The material of all the workpieces used in the experiments is 7075-T6 aluminium and their geometry is prismatic. In the validation experiments, solid blocks are used while in the comparative tests, thin-walled block workpieces with a pre-machined rectangular cavity are used.

A Digital Electronic Automation (DEA) co-ordinate measuring machine is used for inspecting the workpieces. The reported linear accuracy and repeatability from the manufacturer are ±3μm and ±1μm, respectively, with a resolution of 1μm.

6.7 Validation of Dynamic Clamping

Experimental tests are needed to verify the performance of the developed dynamic clamping actuator and its control algorithm. A slot-milling operation is performed on a rectangular block to cut a through slot on its top surface. The workpiece is loaded into the test fixture and secured by only one dynamic clamp applied to its side as shown in Figure 6.19. The clamp is positioned at the same level as the sideways locators and directly opposite to the two secondary locators. The cutting tool is a 10mm four-flute coated HSS straight end mill with a 30° helix angle. No coolant is used during the machining process. The spindle speed is set at 750 rev/min and the feed rate at 100 mm/min. The test fixture loaded with the workpiece is securely attached onto the machine table. As shown in Figure 6.19(b), the machine tool traverses the workpiece along a linear path to produce the slot feature.

(a)

(b)

Figure 6.19 (a) The test fixture loaded with a block workpiece, and (b) the workpiece being machined during a validation test

6.7.1 Validating Experimentation and Discussions

The block workpiece has an overall size of 100(*H*) x 120(*W*) x 220(*L*) mm and it has been designed with a step feature so that the depth of cut is varied from 1 mm to 2 mm (Figure 6.20). Such a feature is deliberately employed so as to provide a surge in the cutting force. The objective is to verify that the dynamic clamp could accommodate the change of the processing force through quickly adapting its clamping intensity.

The clamping intensity for this operation can be generated by the algorithm presented in Chapter 4. The magnitude is 320 N for the initial depth of cut of 1 mm and correspondingly increased to 640 N for the subsequent depth of cut of 2 mm. The function of the control strategy is to regulate the clamping intensity well around the desired level against noise and disturbances and to quickly change the clamping force magnitude as the depth of cut is commanded to alter. For this particular machining operation, the time for the cutter from starting to engage new depth of cut to being fully engaged is estimated to be 3 seconds. This implies that the milling force would jump in this period and the response time of the clamping actuation is preferred to be a fraction of this transient period in order to accommodate any surge in the cutting force.

In the first experiment, the actuator is commanded to increase the clamping intensity from 320 N to 640 N within 1 second (t = 1 s) in a ramp fashion. The actuator starts to change its intensity exactly when the cutter begins to engage the new depth of cut. The actual clamping intensity during the experiment is captured and shown in Figure 6.21. From the force response plot, it is demonstrated that the clamping force is well regulated at the desired force level. During the period of transient depth of cut, there are only slight overshoots and undershoots at the end of the ramp. The response dynamics is quite satisfactory in terms of settling time and oscillation. This indicates the response can even be demanded in a shorter time.

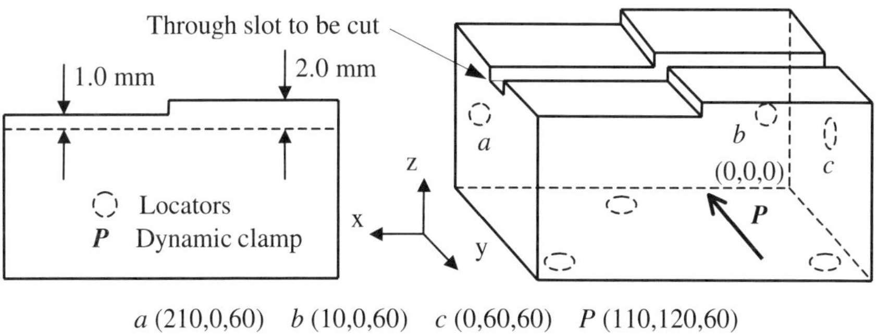

Figure 6.20 The depth of cut variation and the set-up of fixture-workpiece system

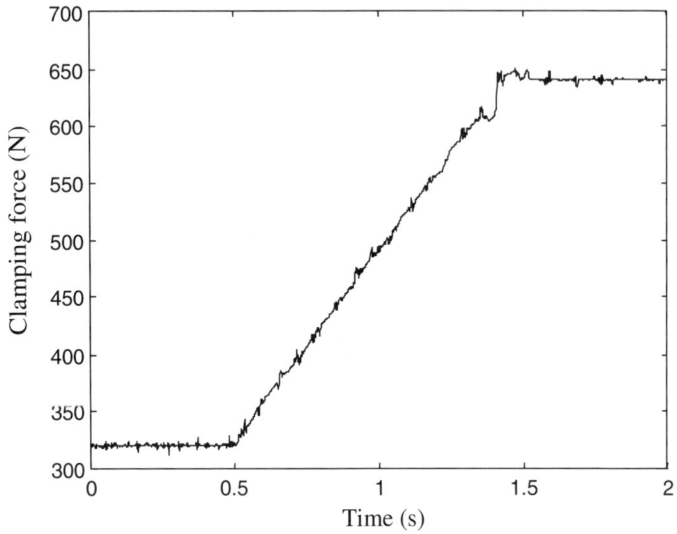

Figure 6.21 Time history of measured clamping force for t = 1 second

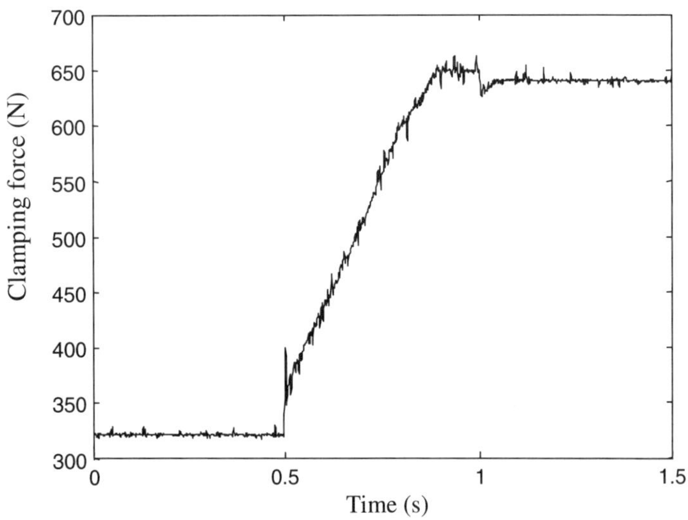

Figure 6.22 Time history of measured clamping force for t = 0.5 second

In the second experiment, the clamping force is commanded to complete alternation within only 0.5 second, or merely one sixth of the transient period. Experiment is repeated as the previous test and the results are shown in Figure 6.22. The regulatory function of the control system is as attractive as the first test. There are reasonable transient overshoots at the beginning of the ramp and response oscillation is slightly increased but still well within moderate limits. Despite the fact that the ramp is much steeper than the first case, the system response is still robust and stable.

The test results indicate that a smooth control output can be achieved by the simplified GMV control strategy, although the algorithm requires the estimation of fewer parameters than a conventional GMV controller. The promising experimental results are very encouraging, confirming that the developed dynamic clamping actuator is stable in providing the required clamping intensity and the control strategy is effective for the clamping force control.

6.7.2 Workpiece Stability under Dynamic Clamping

When a workpiece is being machined during a process, the workpiece stability can be described as equilibrium between all the external and internal forces imparted to it. It is one of the fundamental requirements of a good fixture. Experiments show that a dynamic clamping scheme could help maintain the workpiece stability against the cutting force fluctuation, chatter and vibration.

At the second dynamic clamping test, reaction forces on the three side locators *a*, *b* and *c* (Figure 6.20) are captured by the sensor and data acquisition system and presented in Figure 6.23. During the first half of the tool path, the reactions on locators *a* and *b* vary in a small range (about 30N) with increment on locator *b* and decrement on locator *a*, and locator *c* endures a very stable reaction force. All this indicates that the workpiece is well secured by the dynamic clamp during this period.

When the cutter engages a transitional depth of cut at the mid-point of the tool path, the dynamic clamp quickly adapts its clamping intensity so that the workpiece is still well maintained in stability despite the change in processing force. This is shown in the stable reaction on locator *c* and reasonable variance of reactions on locators *a* and *b*. After that, all the reactions are kept steady throughout the second half of the tool path. It can be concluded that this machining operation is successfully performed and good accuracy can be expected.

In contrast, a conventional fixed clamping scheme may render the workpiece unstable when the cutter encounters variance in depth of cut. Reaction force history from such a fixed workholding scheme is shown in Figure 6.24. During the first half of the path, the workpiece is quite stable and this status can be easily identified by the steadfast reactions on the locators. However, this stability is quickly lost when the cutter engages a new depth of cut.

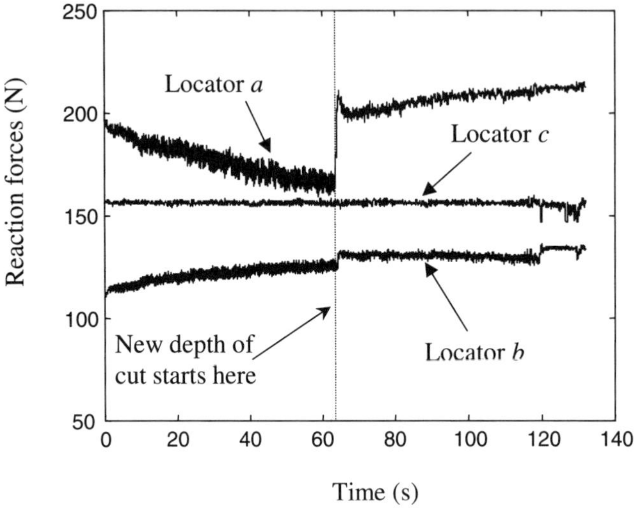

Figure 6.23 A stable workpiece under a dynamic clamping scheme

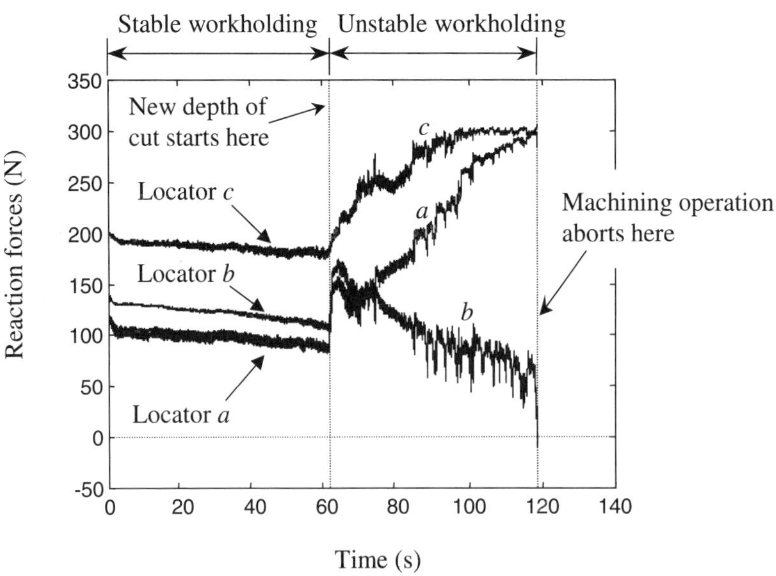

Figure 6.24 An unstable workpiece under a conventional clamping scheme

As shown in the history diagram, the reactions on the locators drastically deviate from the normal conditions. Reactions on locators *a* and *c* quickly increase and reaction on locator *b* alarmingly decreases, indicating the workpiece is being lifted off within the fixture. This is because the applied clamping intensity is no longer sufficient. The lost locating position would jeopardise the geometric accuracy of the workpiece. In addition, owing to the lost contact with base locators, the workpiece starts to vibrate, resulting in large reactions on the locators. Vibration will adversely affect the surface quality of the generated profile. Finally, the milling operation has to be aborted for safety considerations when the reaction on locator *b* detrimentally drops to zero.

A conclusion can be drawn from the results that the workpiece stability can be better achieved with a dynamic clamping scheme than by applying a constant clamping force.

6.8 Impact of Intelligent Fixturing System

6.8.1 Typical Dynamic Workholding Procedure

A typical dynamic workholding procedure commences with the set-up of the intelligent fixture on the machine table. It must be carefully performed before the start-up of a dynamic workholding operation. The set-up consists of several steps. The fixture is first placed on the machine table and the machine tool axes, the cutter and fixture reference frames are aligned. A dial indicator can be attached to the machine spindle so that the indicator can run across the datum surface on the fixture to align the axes. Here it is assumed that the locating datum has already been precisely aligned with respect to the fixture datum surface. The position of the cutting tool relative to the fixture reference frame is then established using an edge finder. Subsequently the NC code is updated so that the reference dimensions in the workpiece and programme reference frames are aligned.

The host computer downloads the optimised fixturing parameters for a part to be machined. The clamp controller then retrieves the dynamic clamping intensity model from the host. After that, the workpiece is loaded into the fixture. Six dial test indicators are brought into contact with the part and their readings are zeroed. Now the host flags the controller to actuate the clamps until individual force set-points are reached. After the workpiece is clamped, the part displacement is determined through the evaluation of the recorded measurements. Tool path compensation is then performed and the transformed NC code is subsequently downloaded to the machine controller. At this time the cutter path is aligned with the actual workpiece location rather than the axes of the machine tool.

The machine controller uses linear or circular interpolations to cut along a prescribed tool path. The clamping controller is simultaneously adapting the clamping intensity with respect to the position of the cutter. At the same time, the acquisition system collects and saves all the force values on the locators and clamps. These actions are continuously carried out until the completion of the machining process.

6.8.2 Experiments on Thin-walled Workpieces

In the aerospace industry, milling operation is widely used for a variety of roughing and finishing processes on aircraft structural components. These components often have thin-walled flanges that are dictated by strength and weight considerations. The main problem with thin-walled workpieces in the machining process is elastic deformation, which can be influenced by many factors such as cutting forces, clamping forces, workpiece rigidity, cutting temperature and internal stresses. Workpiece defection would result in surface error and taper on the finished surface. It is possible to reduce the elastic deformation by applying dynamic clamping forces.

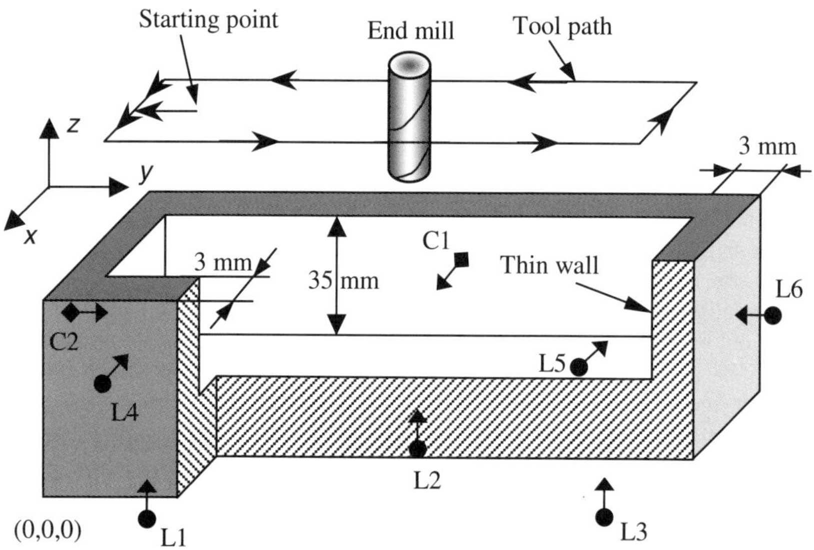

Positions of fixturing elements (mm)

L1	L2	L3	L4	L5	L6	C1	C2
(-10,10,0)	(-110,110,0)	(-10,210,0)	(0,10,55)	(0,210,55)	(-60,220,55)	(-120,110,55)	(-60,0,55)

Figure 6.25 Rectangular pocket milling operation (finishing)

In order to investigate the impact of the dynamic clamping scheme on workpiece quality and the machining process, a finish pocketing operation is performed on a box-shaped thin-walled workpiece. As shown in Figure 6.25, the workpiece has an overall size of 80(*H*) x 130(*W*) x 220(*L*) mm with a rectangular cavity. The wall thickness of the part is to be reduced from 5 mm to 3 mm through finish pocketing by a 20 mm flat-end mill. The depth of cavity is 35 mm so that the pocket walls are milled at 5 mm increments until the bottom of the pocket is reached. Therefore, the axial and radial depths of cut are 5 mm and 2 mm respectively. The spindle speed is set at 500 rev/min and the feed rate at 100 mm/min. Two identical parts are machined with the same cutting parameters following the same machining procedures, but under fixed and dynamic clamping schemes respectively. All the end milling operations are conducted in downward cutting (climb cut) for various reasons such as tool service life, machining accuracy and vibration. The machining process is shown in Figure 6.26 where the part is located by six flat-tipped locating pins according to the 3-2-1 principle and held by two dynamic clamps from its sides. (The schematic fixture configuration is illustrated in Figure 6.18.)

Figure 6.26 Pocketing milling operation in progress under dynamic clamping

It should be noted that the magnitude and direction of the cutting force vary drastically in a pocketing operation. As the cutter reaches a corner of the cavity (Figure 6.27), the engagement width increases sharply from $\boldsymbol{a}_1 = 2$ mm to $\boldsymbol{a}_2 = 16$ mm in this particular case, resulting in much higher cutting force (increased from 264.2N to 482.8N). Once the cutter passes the corner, the cutting force returns to normal but changes its direction. The change in the cutting force is considered in the generation of the dynamic clamping force spectrum.

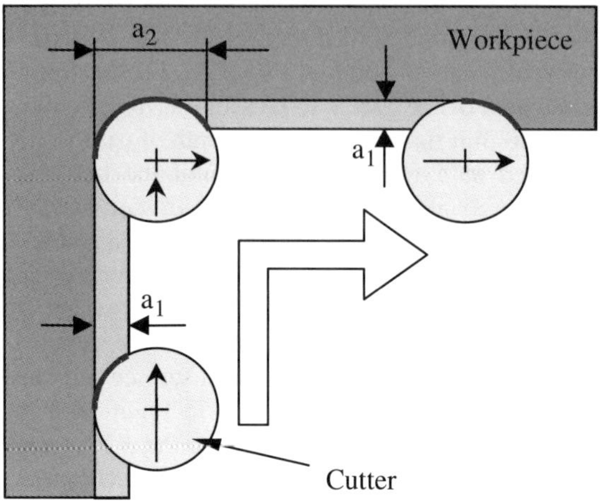

Figure 6.27 Variation of engagement width during cornering

6.8.3 Geometric Accuracy Improvement

In a machining operation on flexible workpieces, both cutter and workpiece deflection can cause errors of the machined surface. Workpiece deflection is induced by the cutting and clamping forces. While the cutting forces produce deflections on both the cutter and the workpiece, such as the case of pocketing thin walls, clamping forces only bring about deformation on the part. For climb milling, the cutter is always deflected away from the workpiece so the final dimension of the part is always larger than desired. The opposite is true for the clamping forces as they usually constrict the part. In this study, it is assumed that the surface errors induced by the cutting forces are the same for both fixed and dynamic clamping cases as the cutting parameters and conditions are identical. Attention is hence concentrated on the surface errors caused by the clamping force-induced deflection.

Surface error is defined as the deviation between the finished surface dimension and the nominal or desired surface that would be produced by a completely rigid machine-fixture-tool-workpiece system. The errors under consideration are macroscopic errors due to workpiece and cutter deflection. Microscopic characteristics of surface roughness due to vibration and chip formation process are not considered.

The profiles of the finished parts can be measured using a co-ordinate measuring machine for determining surface error. For each experiment, the machined surface is inspected several times so as to provide statistically sound measurements. A section view of the thin-walled part is shown in Figure 6.28 where the clamp position is indicated and the measured area on the thin web is 135 mm by 30 mm. The measured area is selected in such a way that the clamping effects can be identified.

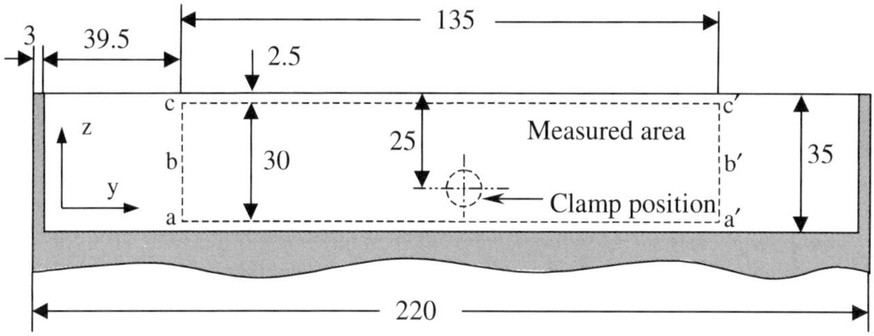

Figure 6.28 Surface error measurements diagram

6.8.3.1 Comparison with Fixed Clamping

As the fixed clamping scheme has to maintain larger clamping forces throughout the machining pass, more material is thus removed around the area opposite the clamping points as the clamping forces cause inward deformation. The finished surface profile is shown in Figure 6.29, where the three curves represent the surface variability along the *y*-direction from the bottom to the top of cut. For each curve, a large negative error (a dip) is generated midway. This dip is caused by the applied clamping force. From the bottom to the top of cut, the profile goes up as the clamping effect diminishes. The overall measured profile is shown in Figure 6.31 where the profile variability is 0.026mm. It can be seen that the surface error is quite large and is not uniform over the finished surface. It is clear that the dominating component of the surface inaccuracy comes from the clamping-induced deflection.

In contrast to the rougher finished surface produced by the fixed clamping scheme, a smoother surface is generated under the dynamic clamping as shown in Figures 6.30 and 6.32 where the profile variability over the measured area is 0.011mm. This is due to the fact that clamping forces are kept relatively low when machining the walls and are only increased correspondingly when milling the four corners with dynamic clamping, and dimension variability is hence significantly reduced. The clamping force is no longer the primary contributor to the surface errors, which are caused by a combination of the part and cutter deflection, cutter run-out, chip formation and vibration. In conclusion, by comparing the profile variability of the machined surfaces, it is established that the machined surface accuracy is notably improved with dynamic clamping.

It is possible to predict the magnitudes of the errors of the machined profiles using the FEM technique. Workpiece deformation analysis can be performed in discrete time instants for the machining operation using the FEM model presented in Chapter 5. This is useful in establishing basic knowledge about the characteristics of the errors produced with different clamping situations. An example of a typical situation is shown in Figure 6.33, with a comparison with experimental measurements, as well as the workpiece deformation distribution under the fixed clamping case in Figure 6.34.

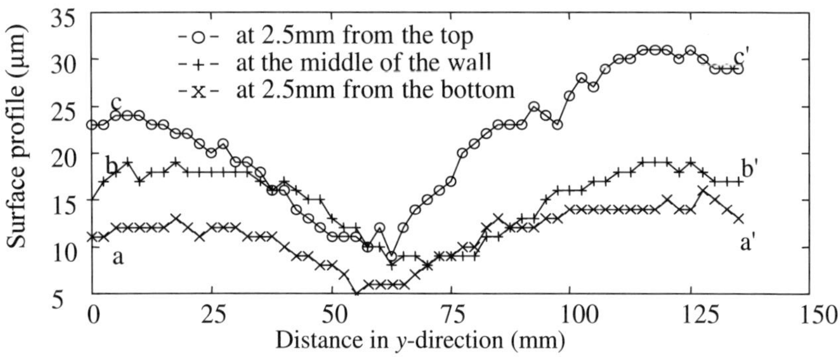

Figure 6.29 Measured surface profile for fixed clamping

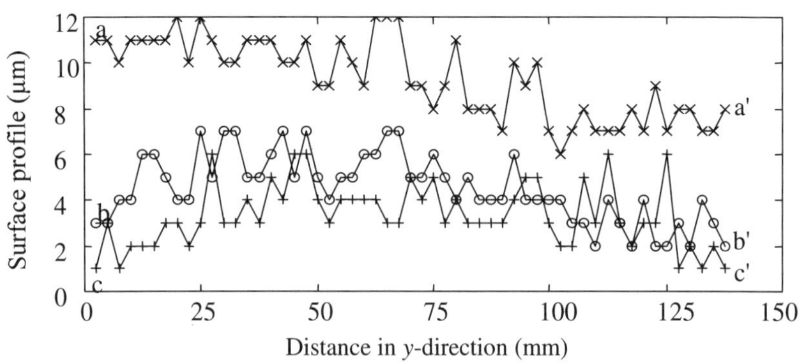

Figure 6.30 Measured surface profile for dynamic clamping

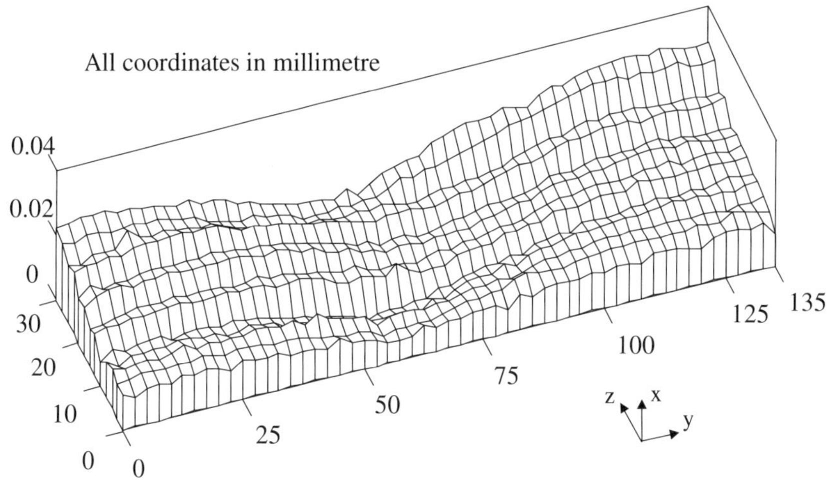

Figure 6.31 Measured profile of milled surface under fixed clamping

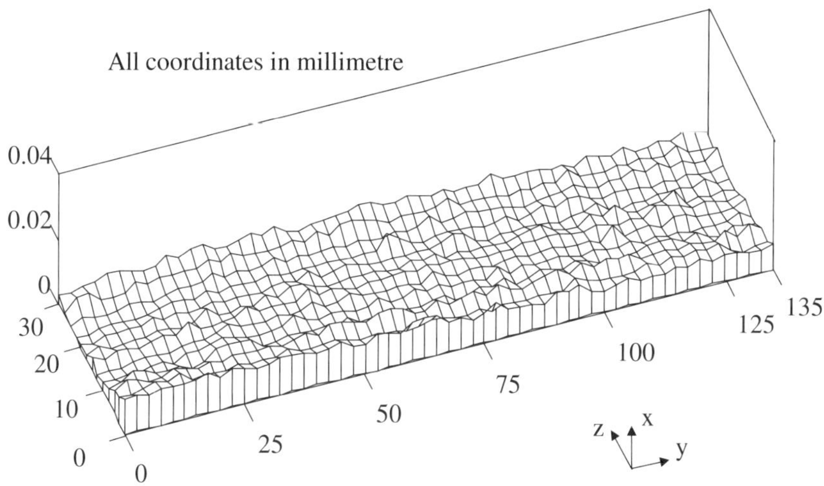

Figure 6.32 Measured profile of milled surface under dynamic clamping

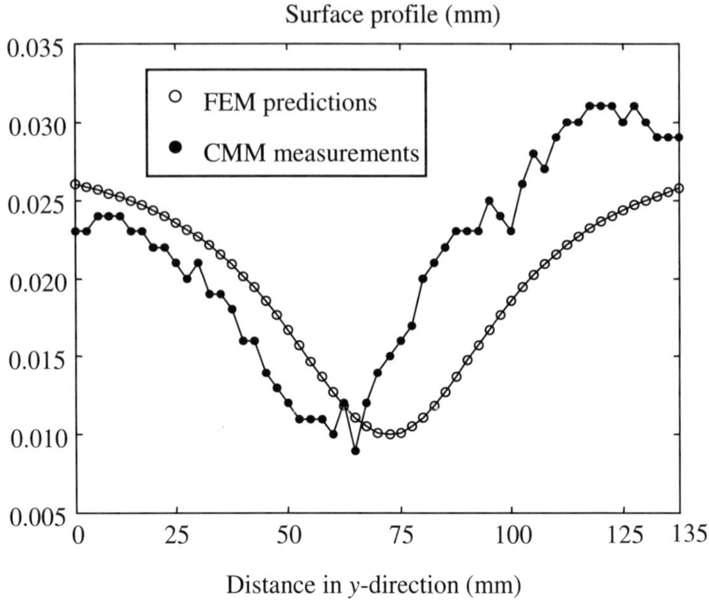

Figure 6.33 Predictions of machined surface profile using FEM (fixed clamping)

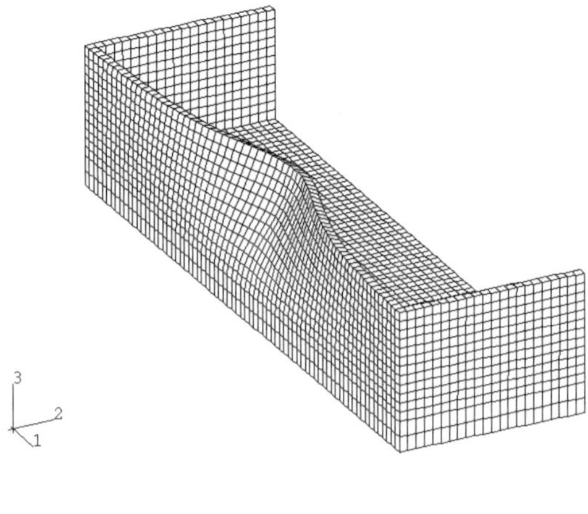

Figure 6.34 Typical deformation of the workpiece under the fixed clamping scheme

6.8.4 Finished Surface Quality

During the slot-milling operation in the validation tests, the bottom surface of the slot is generated by the flat end of the mill which is perpendicular to the machined surface. The appearance of the finished surface is shown in Figure 6.35, where the tooth mark spacing is the feed per revolution (0.133 mm) because the cutter is always deflected away from the part opposite to the feeding direction so that the tilted cutter leaves one deeper groove mark for each revolution. As the workpiece is held stable, the tooth mark pattern is regular and clean-cut, resulting in improved surface quality.

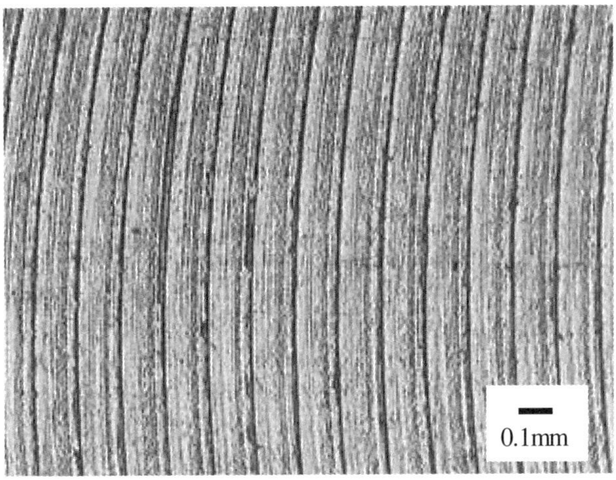

Figure 6.35 Photograph of machined surface under stable workholding

In contrast, unwanted workpiece instability in a fixture can result in both excessive geometric error and marring of the machined surface. One source of instability is the under-clamping actuation, which would result in severe workpiece slippage, displacement and vibration, thus spoiling the surface quality. Figure 6.36 shows the generated surface under unstable workholding conditions. Irregular marks are introduced into the machined surface by part instability, which degrades the surface finish. Therefore, workpiece stability achieved in the dynamic clamping scheme has improved the surface quality.

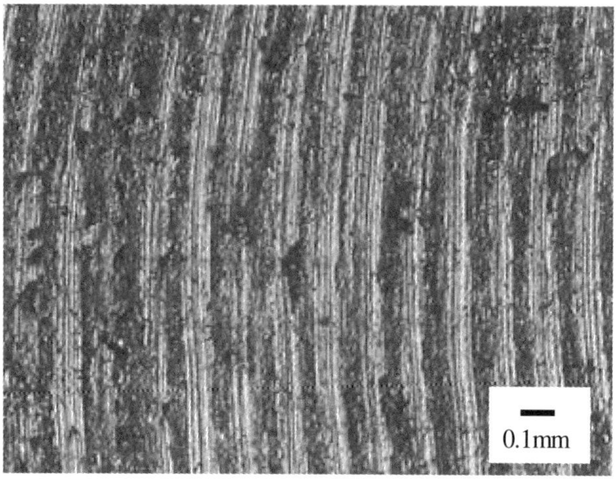

Figure 6.36 Photograph of machined surface under unstable workholding

6.8.5 Process Cycle Time

When milling with small engagement widths in machining the sides of a pocket, the recommended feed/tooth values are often quite high in order to achieve the correct chip thickness. The feed rate is usually too high for the cutter when it goes around a corner. In order to avoid large deflections and vibrations induced by gouging when milling a corner, the feed rate has to be greatly reduced in a conventional fixture. This implies that the process cycle time will be prolonged. However, in the intelligent fixturing system with a dynamic clamping function, such slowing down of feed rate is unnecessary as the adaptive clamping forces will achieve a smooth operation. Both roughing and finishing operations can also be completed in one operation pass. Process cycle time can hence be reduced and productivity improvement is to be expected by employing a dynamic clamping scheme.

6.9 Summary

Using existing and emerging technologies in the areas of process planning and computer control, an intelligent fixturing system for precision machining operations has been developed. This system can control fixturing operations in concurrence with machining operations.

In this chapter, the architecture and functions of the intelligent fixturing system are presented. Dynamic clamping provides fast clamping action and consistent clamping forces. The sensory feedback of the real-time actual clamping forces in the system is particularly significant in an automated manufacturing environment. It is now possible for a machinist to exert a precise clamping intensity, which is highly desirable in maintaining good machining accuracy. The presence or absence of a sensory signal may also be used as an initial indication of fixturing system malfunctions, e.g., clamp failure or clamp misplacement.

The developed intelligent fixturing system has proven its usefulness for workpiece quality improvement and productivity enhancement. It is envisaged that this system would eventually lead to a new form of workholding device that can inspect the workpiece location, correct the part displacement through tool path compensation, contain the workpiece distortion through dynamic clamping, adapt the fixturing parameters based on sensory feedback, monitor and self-regulate the workholding process, and is consequently suitable for an automated manufacturing environment.

6.10 Bibliography

Asada, H. and A.B. By. Kinematic Analysis of Workpart Fixturing for Flexible Assembly with Automatically Reconfigurable Fixtures. *IEEE Journal of Robotics and Automation*, 1(2), pp. 86-94. 1985.

Ceglarek, D. and J. Shi. Fixture Failure Diagnosis for Autobody Assembly Using Pattern Recognition. *Journal of Engineering for Industry*, 118(1), pp. 55-66. 1996.

Chan, K.C. and C.S. Lin. Development of a Computer Numerical Control (CNC) Modular Fixture - Machine Design of a Standard Multifinger Module. *International Journal of Advanced Manufacturing Technology*, 11(1), pp.18-26. 1996.

Chan, K.C., B. Benhabib and M.Q. Dai. A Reconfigurable Fixturing System for Robotic Assembly. *Journal of Manufacturing Systems*, 9(3), pp.206-221. 1990.

Chou, Y.-C., V. Chandru and M.M. Barash. A Mathematical Approach to Automatic Configuration of Machining Fixtures: Analysis and Synthesis. *Journal of Engineering for Industry*, 111(4), pp.579-584. 1989.

Cogun, C. The Importance of the Application Sequence of Clamping Forces on Workpiece Accuracy. *Journal of Engineering for Industry*, 114, pp.539-543. 1992.

Deckel Maho GmbH. *AxTra Operating Manual*. 1996.

DeMeter, E.C. Restraint Analysis of Fixtures which Rely on Surface Contact. *Journal of Engineering for Industry*, 116(2), pp. 207-215. 1994.

Gandhi, M.V. and B.S. Thompson. Phase-change Fixture for Flexible Manufacturing Systems. *Journal of Manufacturing Systems*, 4(1), pp. 29-39. 1985.

Gupta, S., A. Bagchi and R. Lewis. Sensor Based Fixturing System. In *Recent Developments in Production Research*, ed. by A. Mital, pp.11-16. Elsevier. 1988.

Hazen, F.B. and P. Wright. Autonomous Fixture Loading by a Machine Tool. *Proc. of ASME Symp. on Manuf. Sys.*, pp. 105-111. 1988.

Helsel, R. *Cutting Your Test Development Time with HP VEE: an Iconic Programming Language*. PTR Prentice Hall. 1994.

Hockenberger, M.J. and E.C. DeMeter. The Effect of Machining Fixture Design Parameters on Workpiece Displacement. *Manufacturing Review*, 8(1), pp.22-32. 1994.

Huang, Y. and T. Hoshi. Improvement of Flatness Error in Milling Plate-shaped Workpiece by Application of Side-clamping Force. *Precision Engineering*, Vol.24, No.4, pp.364-370, 2000.

Jeng, S.L., L.G. Chen and W.H. Chieng. Analysis of Minimum Clamping Force. *Int. J. Mach. Tools Manufact.*, 35(9), pp.1213-1224.1995.

Kuo, B.C. and D.C. Hanselman. *MATLAB Tools for Control System Analysis and Design*. Prentice Hall. 1994.

Kurz, K., K. Craig, B. Wolf and F. Stolfi. Developing a Flexible Automated Fixturing Device. *Mechanical Engineering*, 1994-7, pp. 59-63. 1994.

Lee, J.D. and L.S. Haynes. Finite-Element Analysis of Flexible Fixturing System. *Journal of Engineering for Industry*, 109(2), pp.134-139. 1987.

Lee, S.H. and M.R. Cutkosky. Fixturing Planning with Friction. *Journal of Engineering for Industry*, 113, pp. 320-327. 1991.

Lu, S.S. and J.L. Chu. Development of a Novel Co-ordinate Synchronising Fixture System. *Proc. Japan/USA Symposium on Flexible Automation*, pp. 863-869. 1996.

Nee, A.Y.C., K. Whybrew and A. Senthil kumar. *Advanced Fixture Design for FMS*. London: Springer-Verlag. 1995.

Nnaji, B.O., S. Alladin and P. Lyu. A Framework for a Rule-base Expert Fixturing System for Face Milling Planar Surface on a CAD System Using Flexible Fixtures. *Journal of Manufacturing System*, 7(3), pp.193-207. 1988.

Raghu, A. and S.N. Melkote. Analysis of the Effects of Fixture Clamping Sequence on Part Location Errors. *International Journal of Machine Tools and Manufacture*, Vol.44, No.4, pp.373-382. 2004.

Sela, M.N., O. Gaudry, E. Dombre and B. Benhabib. Reconfigurable Modular Fixturing System for Thin-walled Flexible Objects. *International Journal of Advanced Manufacturing Technology*, 13(9), pp. 611-617. 1997.

Shirinzadeh, B. Issues in the Design of the Reconfigurable Fixture Modules for Robotic Assembly. *Journal of Manufacturing System*, 12(1), pp.1-14. 1993.

Trappey, A.J.C. and C.R. Liu. Issues of automatic fixture design in computer integrated manufacturing system. *Proceedings of IASTED International Conference on Robotics and Automation*, pp.149-154. 1988.

Wu, N.H., K.C. Chan and S.S. Leong. Fixture Verification Based on The Analysis of Multi-Discipline Frictional Contacts. *Manufacturing Science and Engineering ASME*, Vol. 3-1, pp.735-745. 1995.

Chapter 7

A Database Management System for Modular Fixtures

7.1 Introduction

Modular fixtures have gained popularity in the manufacturing industries, due largely to the shorter time-to-market requirement, higher variety and smaller batch-size products. Modular fixtures can be assembled/disassembled easily, eliminating long construction time, allowing reuse and reduction of storage requirements.

One of the associated problems with the use of modular fixtures is their proper inventory management. Some of the considerations are providing sufficient quantity to meet operation needs, setting replacement policy for worn-out elements, and tracking the use of the elements in ongoing assignments. A well-designed database management system would be a useful tool in handling this problem. Information such as utilization status and inventory level of fixture elements, the weight and cost of a fixture design can be easily extracted.

Extensive research has been done in modular fixture design synthesis and analysis using computer tools (Huang and Trappey, 1992; Dai *et al.*, 1997; Fuh *et al.*, 1995; Nee *et al.*, 1995), however, little has been reported in the area of inventory management of modular fixture elements.

This chapter presents a Fixture Element Database Management System (FE_DBMS) developed in C language together with some of its library of functions on a UNIX platform. This is to facilitate integration with UNIX-based CAD tools and to allow real-time inquiry of the availability of fixturing elements while performing a part design.

7.2 System Overview

The Fixture Element Database Management System (FE_DBMS) developed is not a mere data collection tool, it can be used to perform simulation of job shop activities and report on the utilization status of the fixture elements. In addition, the management could decide, from the available data, how best they should schedule the jobs so that a minimum amount of fixture elements need to be held in the inventory.

There are five levels of requirements as described in the following five modules.

7.2.1 Enquiry Module

The system is able to generate, upon inquiry, the following useful information:

- Minimum inventory level for the proper functioning of a job shop,
- Useful lifespan of elements expended,
- Number of elements used per month,
- Weight of a fixture design (without the workpiece),
- Cost of a fixture design (without the workpiece, based on fixture element cost),
- Percentage of each category of elements used in a job, and hence the utilization of a particular class of elements.

7.2.2 Element Database Module

For the simulation runs, the number of fixture elements is assumed to be unlimited. In other words, the jobs can draw as many similar elements as possible without fear of running out from the inventory. At this stage, constraint of elements and suggestion for selecting alternative elements are not considered.

7.2.3 Job Module

The system handles information such as the weight and cost of fixture elements involved and the commencement/completion dates of the job. The system also checks the validity of the part number of elements used for each job.

7.2.4 Schedule Module

For every job sent for machining, there are specific commencement/completion dates that have to be adhered to. The completion date determines the priority of the job. The earlier this date is, the higher will be the priority. The system would allow the scheduling of jobs to be prioritised in this way, however, the system provides some flexibility in setting priority as desired by the user.

7.2.5 Simulation Module

A simulation of the job shop activities throughout a calendar year, or any period, can be performed in this program.

During program simulation, it detects the commencement and completion of jobs based on the schedule of these jobs. Elements needed for these jobs will be quantified through this simulation. Information from this run is stored in a different location from that of the Element Database, as it is more of a "safety-level" database (i.e., minimum inventory level).

7.3 System Implementation

In system implementation, a modular approach was adopted. Each module developed was subdivided into several sub-modules. The five modules mentioned in Section 7.2 were developed with this approach.

Figure 7.1 shows the level of integration/interaction between the modules within FE_DBMS.

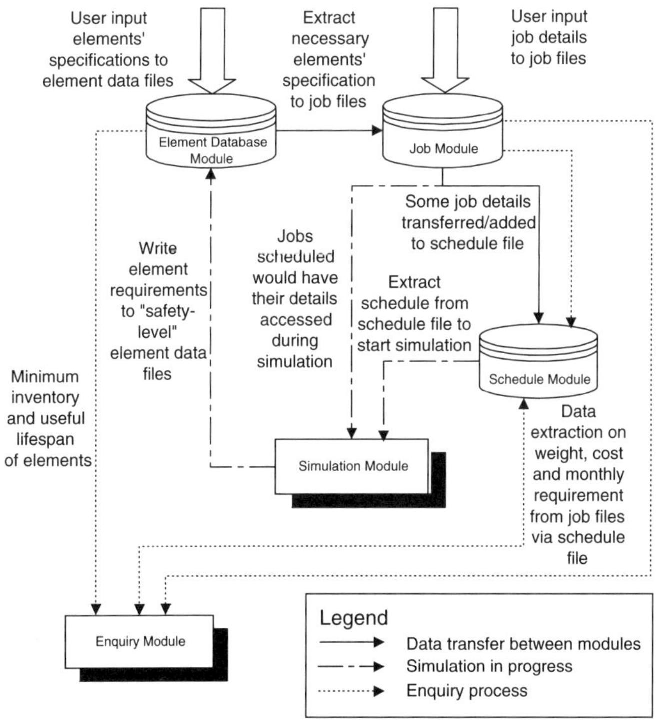

Figure 7.1 Overview of integration/interaction between modules in FE_DBMS

7.3.1 Data Storage Capability

Data storage capability of the FE_DBMS is supported by the files created in the Element Database, Job and Schedule modules. The interconnectivity of modules in the FE_DBMS can be demonstrated by the data transferring activities occurring between these three modules.

The files used in each module are initialised with a pre-defined/user-defined amount of data storage space. Additional data space may be added at a later stage. For Element Database and Job modules, an option to add new attribute of interest to the files for recording a new type of data is available. These files are of random access type to maintain the integrity of the data stored.

In the Element Database module, the element data files created contain specifications of the fixture elements found in the inventory. Specifically, six data files, corresponding to the six categories of fixture elements, are to be initialised. Each of these six files has a corresponding "safety-level" element data file being initialised. This latter set of files is useful to the Simulation module for the recording of the derived information on completion of a simulation of job shop activities, while the former set receives direct input from a user on the types of fixtures elements available and their specifications. Table 7.1 shows what is typically stored in an element (stop) data file.

Table 7.1 Types of stop element in database

S/no	Part No	Status	Wt (kg)	Cost/pc ($)
1	20112022	1	0.33	111.7
2	20212022	1	0.31	114.2
3	20216180	1	0	88
4	21012050	1	0.5	30
5	21016063	1	1.2	32
6	21112001	1	0.36	24.1
7	23012080	1	4.6	45
8	23012163	1	3.9	46
9	23016063	1	3.7	42

Note: '0' value under the '*Wt*' column depicts a lack of information at the time of compilation

Table 7.2 shows what is typically stored in a "safety-level" element (base) data file.

Table 7.2 Base elements required upon completion of simulation run of the five jobs

S/no	Part No	Element No	Status	End_dd	Usage Level	Hour Used
1	010608012	2	1	0 0 0	1	672
2	010608012	1	1	0 0 0	1	672
3	040404012	2	1	0 0 0	1	244.8
4	040404012	1	1	0 0 0	2	2443.2
5	040808016	1	1	0 0 0	1	748.8

In the Job module, job files are created to contain job details. Such details include the part number of fixture elements used, and the commencement and completion dates that are provided by users. Technical specifications such as the weight and the cost of these elements are transferred from element data files to the job file. A typical job file (`j3`) contains the information found in Table 7.3. Figure 1.5 of Chapter 1 is the fixture designed for this job.

Table 7.3 Part list of a typical job file `j3`

Job name:	j3	
Filename:	j3.dat	
Workpiece: Bracket	**Project Name:** May_Jul	**Commence Date:** 15/5/1999 **Completion Date:** 5/7/1999
S/no	**Part Number**	**Quantity**
1.	040404012	1
2.	12112120	3
3.	21012050	3
4.	33012070	3
5.	60012065	3

In the Schedule module, a schedule file is created to contain details of scheduled job shop activities. This schedule file is updated whenever a new job file is created and needs to be added to the job shop schedule. One important set of data being transferred from the job file to the schedule file is the commencement and completion dates of a job. The set of data contained in the schedule file is crucial to the simulation of the job shop activities.

The data storage capability of the FE_DBMS is summarized in Figure 7.2.

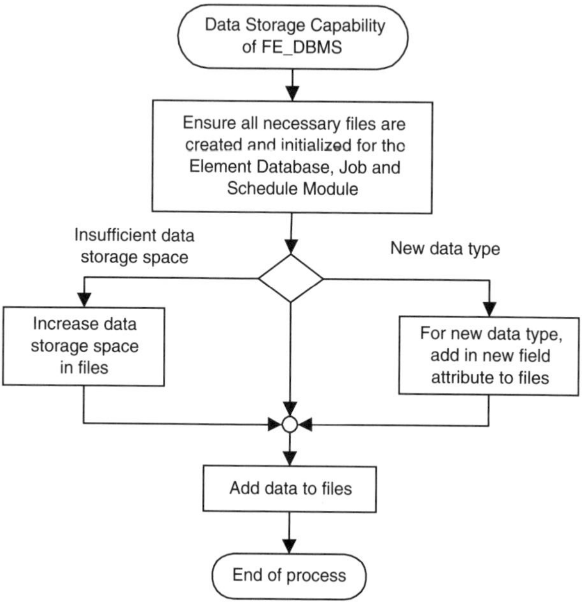

Figure 7.2 Flowchart for data storage capability of FE_DBMS

7.3.2 Data Access Capability

There are three levels of data access to the data that are stored in the files of the FE_DBMS. The first level allows for read-only access. The second and third level allows for read-write access. Such a level of access allows for amendment or complete removal of any set of data. These three levels of access are available to the Element Database, Job and Schedule modules (Figure 7.3).

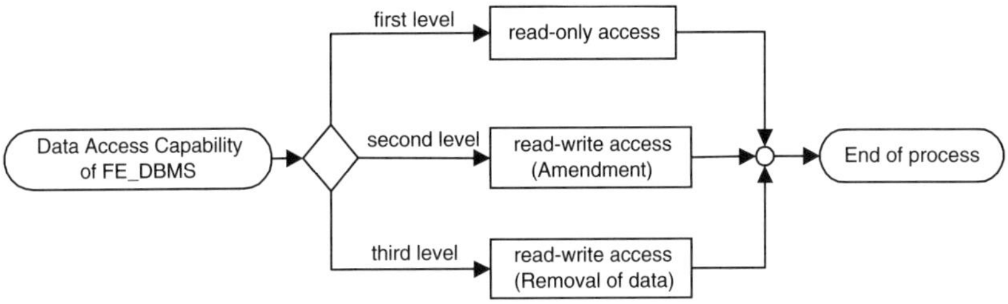

Figure 7.3 Flowchart for data access capability of FE_DBMS

7.3.3 Data Sorting Capability

This is unique to the Schedule module of the FE_DBMS. Such a capability can be interpreted as another level of data access. There are two levels of data sorting capability. By default, job priorities in the schedule file are sorted by their completion dates. The earlier the completion date, the higher is the job priority in the schedule ranking. At this level, no user interference is allowed. At another level, a user could actually determine a job's priority in the schedule, which cannot be overwritten by the default level of data sorting (Figure 7.4).

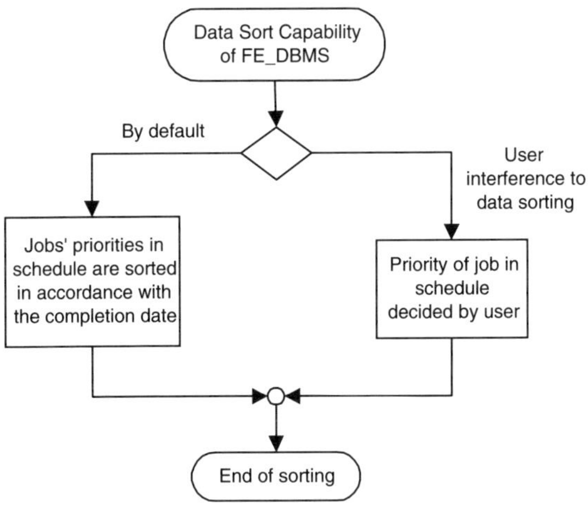

Figure 7.4 Flowchart for data sorting capability of FE_DBMS

7.3.4 Simulation Capability

The Simulation module of the FE_DBMS provides the simulation capability. It will essentially generate two sets of important information upon a successful simulation of the job shop activities. One set is the minimum inventory level for the proper functioning of the job shop, and the other is the expended lifespan of all the elements used in the jobs.

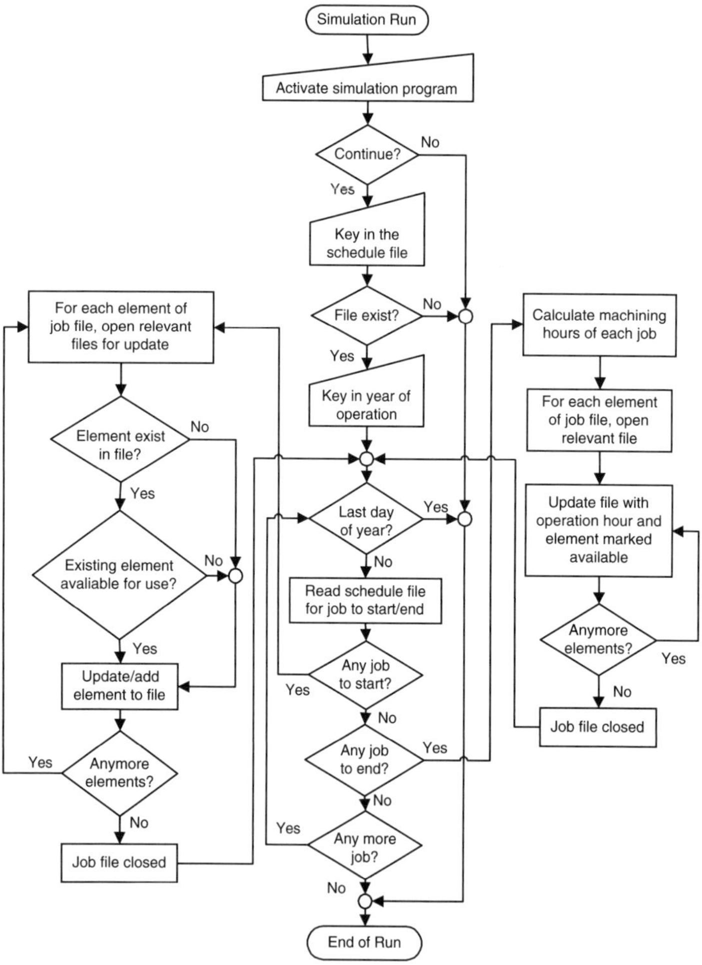

Figure 7.5 Flowchart for implementation of simulation program

Two sets of data are crucial to the success of the simulation run. The first set is that of the data found in the schedule file. With this set of data, the Simulation module can identify the status of a job and determine which job will commence/complete. A commencing job has higher priority to a completed job. For each commencing job, its job file is accessed to determine the fixture elements needed. This is the second set of important data.

During a simulation, the fixture elements needed for the commencement of a job are updated to the one of the six "safety-level" element data files based on their categories. This updating process includes checking for any available fixture elements in the inventory that can meet the demand of this job. If available, they will be reused. Otherwise, new fixture elements are "drawn" out to meet the demand. In this way, the amount of fixture elements found in these files reflects the minimum amount the inventory would need to hold for the proper functioning of the job shop. On completion of a job, the number of operation hours gone through by the fixture elements is updated to these files as well. Figure 7.5 is a flowchart of the simulation process.

7.3.5 Data Extraction Capability

In FE_DBMS, there are a lot of embedded data within each module. The Enquiry module permits the extraction of such data. There are six focus areas. Extracted data are stored in the sequential access file format. This type of file format is supported by SQL-based programs, and such programs can provide graphical presentation of the data extracted. Generally, the process of enquiry can be summarized in Figure 7.6. The six areas of focus of the Enquiry module are:

7.3.5.1 Minimum Inventory Level

This set of data is stored in the six "safety-level" element data files. The available information reveals the types of element used of a particular category of element, together with their quantity. This quantity is the minimum number of each element needed to ensure proper functioning of the job shop.

7.3.5.2 Useful Lifespan of Elements Expended

Another set of information embedded in the "safety-level" element data files is that of the number of operation hours each element has gone through. When an enquiry is made, a user will have to indicate the percentage value of useful lifespan expended. This value is compared to the ratio of operation hours to the estimated total useful lifespan. Any element having this ratio higher than the input value by a user will have its part number displayed.

7.3.5.3 Number of Elements Used Per Month

With this enquiry, the level of inventory usage for each month can be shown. This enquiry is made on a monthly basis. The schedule file will be accessed to determine which jobs are active during the month of enquiry. These jobs will have their job files accessed to evaluate the status of the inventory level of the elements used.

7.3.5.4 Weight of Fixture Design (Without Workpiece)

This enquiry provides information on the weight of the designed fixture of each job above a particular level, as indicated by a user. It is simply a summation of the weights of all the elements used in a job. All the scheduled jobs will be evaluated for their weight.

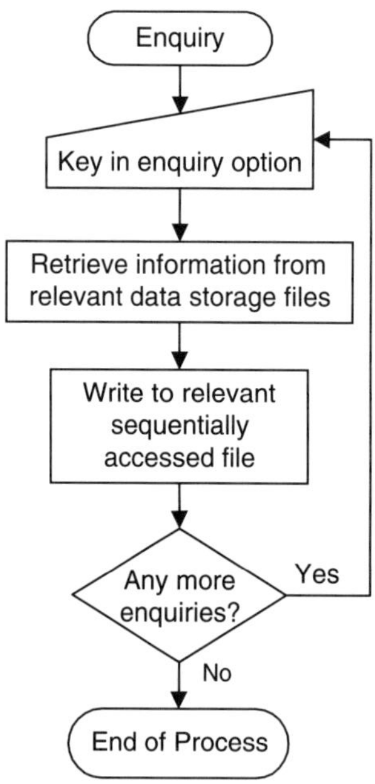

Figure 7.6 Flowchart for general implementation of enquiry processes

7.3.5.5 Cost of Fixture Design (Without Workpiece)

This enquiry is similar to that of the weight enquiry, except the basis of comparison is the cost of the designed fixture. It must be noted that the cost is a simple summation of the costs of each element in the job.

7.3.5.6 Percentage of Each of the Category of Element in a Job

In this enquiry, the number of each type of element used in a job is noted. A user determines which particular job file to enquire about.

7.4 Evaluation of the FE_DBMS

7.4.1 Inventory Enquiry

In any job shop environment, it would have the inventory list of all its fixture elements. A simulation run on its forecast of jobs will give an evaluation on the minimum inventory holding of elements required by those jobs. A comparison between these two sets of information would illustrate if there is sufficient amount of elements for the proper functioning of the job shop (Figure 7.7).

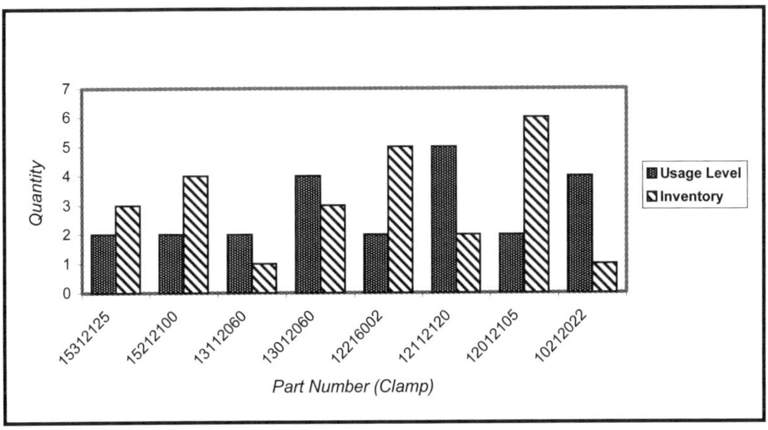

Figure 7.7 Usage level vs. inventory level chart

As the simulation is based on a schedule file, re-scheduling of the job shop activities may provide varying results that could provide the management with some alternative solutions.

7.4.2 Cost Enquiry

With the enquiry on the cost of fixture designs made possible, similar designs should not differ too much from this parameter. However, if there are big differences, such an enquiry would allow the management to focus on fixture designs that are too costly (Figure 7.8).

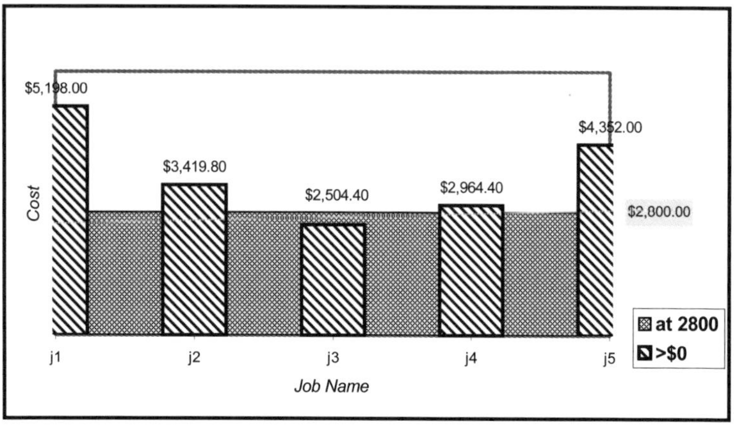

Figure 7.8 Cost of fixture design (fixture elements only) chart

7.4.3 Weight Enquiry

With the enquiry on the weight of fixture designs made possible, the management would have a rough idea on what sort of material handling equipment to invest in (Figure 7.9).

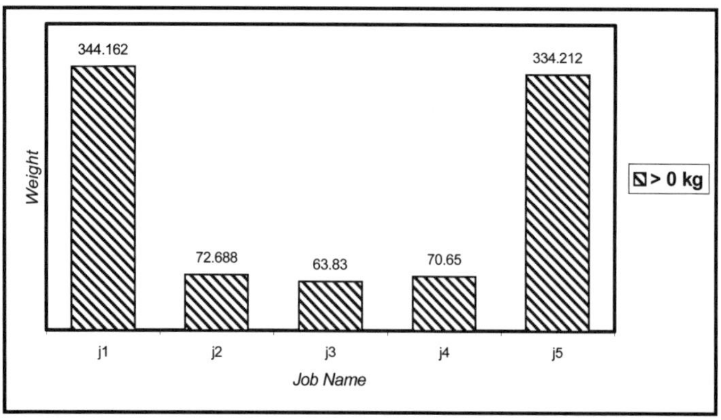

Figure 7.9 Weight of fixture design > 0 kg (without workpiece) chart

7.4.4 Monthly Enquiry

With the monthly enquiry, the level of element usage is reflected. If in a particular month, the usage level is especially high, the management may decide to reschedule some jobs to lower it down (Figure 7.10).

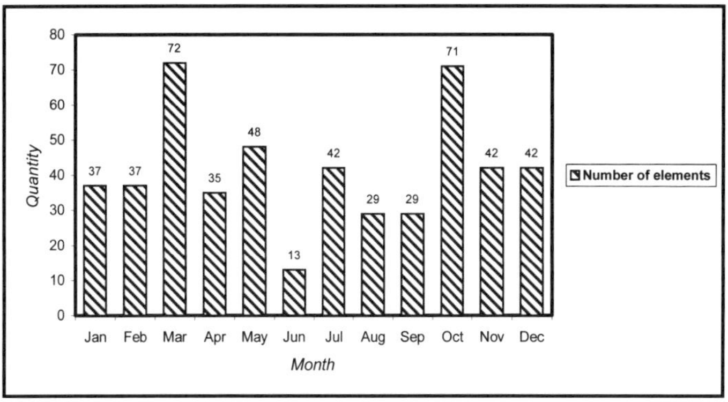

Figure 7.10 Monthly usage of fixture elements chart

7.4.5 Purchasing Policy

In deciding on purchasing or replacement policy, the FE_DBMS has also demonstrated that it can provide management with relevant information. The level at which an element is deemed to be still useful can be decided by the management, and using this system, those which failed to meet this criterion can be highlighted and subsequently replaced (Figure 7.11).

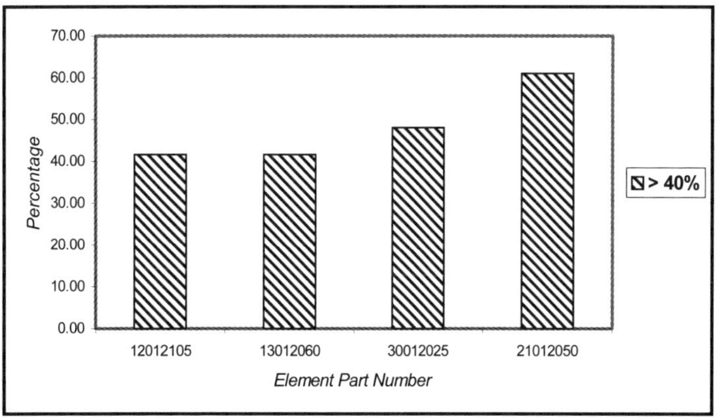

Figure 7.11 Elements with > 40% of the lifespan expended chart

7.5 Summary

Using the Simulation module, together all the data storage capability of the Element Database, Job and Schedule modules, the FE_DBMS has provided useful and relevant information through the Enquiry module.

The Element Database module has fulfilled its role of containing the specifications of all the available types of elements. Since unlimited resource of fixture elements is assumed, the quantity of each element will not be stored. However, in another portion of this module, it has the capability to store the quantity of elements used after a simulation run. The Job module allows the creation of job files. These job files, besides containing their fixture element requirements together with the individual element specifications, also store the details of the commencement and completion dates of the jobs.

The Enquiry module draws on the specifications from these job files to furnish enquiry on weight and cost of a fixture design. The Schedule module draws on the details of commencement and completion dates of each job for handling scheduling activities. The success of a simulation run depends on the information from the Job and Schedule modules. Simulation results on unknown parameters like minimum inventory holding and useful lifespan left of each element provide useful management information.

In order to verify that the five modules have worked well to fulfil the system's requirement, a case study on a simulation of five jobs was done. Not only was the system able to generate the accurate information on the minimum inventory holding, other information like the total weight and cost of fixture designs were also accurately evaluated.

7.6 Bibliography

Dai, J.R., A.Y.C. Nee, J.Y.H. Fuh and A. Senthil kumar. An Approach to Automating Modular Fixture Design and Assembly. *Proc Instn Mech Engrs Part B, J. of Engg. Manuf.*, Vol. 211, pp.509-521. 1997.

Fuh, J.Y.H., A.Y.C. Nee, A. Senthil kumar and J.C.S. Teo. IFDA: An Interactive Fixture Design and Assembly Environment. *Int. J. of Computer Applications in Technology*, 8(1/2), pp.30-40. 1995.

Huang, S.H. and A.J.C. Trappey. The Integration of Modular Fixture Database, Fixturing Knowledge Base and 3-D Fixture Planning Interface. *Computers & Industrial Engineering*, Vol.23, No.1-4, pp.381-384. 1992.

IMAO. *IMAO Venlic Block Jig System (BJS) Fixturing Guide*. IMAO Corporation.

Nee, A.Y.C., K. Whybrew and A. Senthil kumar. *Advanced Fixture Design for FMS*. London: Springer-Verlag. 1995.

Chapter 8

An Internet-Enabled Smart Interactive Fixture Design System

8.1 Introduction

With the advent of Internet technology, many businesses have adopted the Internet as a communication platform or even port some of their core operations onto it. This has enabled them to perform almost instantaneous transactions with their partners worldwide. The manufacturing community should also adopt this approach in their operations, including fixture design. This is especially crucial with the push in the industry to implement distributed manufacturing and to offer rapid manufacturing capabilities (Smith and Wright, 1996). An Internet-enabled system for fixture design would not only save costs and time, it would also help promote new levels of collaboration between designers and their counterparts, as well as with clients, regardless of geographical constraints. The ultimate aim of the proposed system would be to allow users economical and easy access to fixture design functions, and be able to work on fixture designs or to present solutions to their clients even while they are on the move, utilizing any forms of computing platforms.

8.1.1 Standalone Systems

A number of semi-automated and automated fixture design systems have already been developed and implemented on the CAD/CAM systems. Such systems provide fixture design functions such as a parametric fixture element database (Yap, 1999) and built-in interference checking routines to ensure collision-free design solutions (Dai *et al.*, 1997) and to provide for cutter path allowances (Senthil kumar, 1999). These systems are all based on standalone machines or closed license based networks. One of the greatest drawbacks of these systems is the extremely high capital investment that would have to be incurred in order to set up the commercial modelling packages needed to drive them. Another problem faced is the issue of having to constantly maintain and upgrade as newer versions of the fixture design systems and the CAD/CAM platform are being developed and introduced. The proposed system would provide an alternative to this and offer fixture design functions at a small fraction of the costs needed to implement these systems. It would also remove the need for users to constantly perform upgrades.

8.1.2 Internet-Enabled System

The Internet has expanded rapidly over the last few years and will continue to do so in the foreseeable future. It has grown from just a means of posting and tracking articles and data to a rich medium of exchange that offers a multitude of services. As Internet infrastructure and networking bandwidth continue to improve with advancement in technology, it makes sense to tap the potential of this rich resource in the field of manufacturing. Well-developed Internet-enabled manufacturing applications would bring about huge time and cost savings by harnessing the high speed communication and mass accessible features of the Internet.

As with any Internet-based service, the new fixture design system will have a very wide end-user reach, it can be offered at a very low cost compared to systems implemented on license-based commercial modelling packages. It would not incur any set-up costs on end users. They will be able to download an enabling client program that can access the fixture design functionalities hosted on server machines and be charged only when they make use of the program. Many potential end-users are likely to prefer such an option as opposed to having to dedicate large amount of resources in a costly dedicated system. This opens up a new segment of the manufacturing community to the automated fixture design solutions.

8.1.3 Java and Java3D

FixtureNet offers simple fixturing solutions accessible over the Internet using a Web browser (Wagner *et al.*, 1996). This system presents several limitations. Interaction with the fixture design service is not real-time and the time taken to return solutions is quite substantial. Being browser based, the user interface is also largely constrained by hypertext markup language (HTML). Visualization of the workpiece and fixture elements is also confined to 2D layouts. Hence it is proposed to offer real-time 3D fixture design functions on the Internet. Built upon a prototype solid modelling interface, the system presents users with the options of loading a predefined workpiece to utilize the fixturing functions or to model the workpiece using the modelling functions of the solid modelling interface itself. This would be a more true-to-life system design. Another important feature of the new system would be its multi-client support. Two or more users would be able to simultaneously work on the same workpiece from different remote locations.

Basic interactive fixture design (IFD) functions across the Internet have been developed. The user will have to make most of the decision as to the choice and positioning of the fixture elements. The main objective of this initial version is to demonstrate the feasibility of true 3D fixture design remotely over the Internet implementing client-server architecture. Further improvements are being made and a simple semi-automated fixture (SFD) has been developed. Improvements are being made to develop a fully automated fixture design (AFD) functions in order to achieve a truly comprehensive system.

8.2 System Architecture

The detailed schematic of the complete system is shown in Figure 8.1

8.2.1 Server Side

The server end of the system contains a few main portions. These include a fixture element database, the modelling kernel, server classes that drive the solid modelling and fixture element database, the Remote Method Invocation (RMI) interface and finally, an Apache Web server.

The server classes consist of the modelling classes and the feature modelling classes. Modelling classes contain the functions for the creation of the basic primitives such as blocks, cylinders, cones, etc. Feature modelling classes handle the modelling of the various features provided in feature library. Both are interfaced with the modelling kernel using a Java Native Interface (JNI). When the client makes a function call to the server through the RMI interface, the server classes will relay the call to the kernel, which will then perform the necessary modelling tasks.

The fixture element database contains a whole array of fixture elements in Parasolid format files (.xmt_txt). Theses include baseplates, support pins, locating cylinders, stops, clamps and risers from IMAO Corporation's Venlic Block Jig System (BJS) range of elements. The kernel uses these files to generate the Facet Data (FD) XML files that would be used for the rendering and visualization purposes by the client.

The Apache Web server has the function of hosting and transmitting the two types of XML files shown in Figure 8.1. The (FD) XML files reside on the Web server and are retrieved by the client. The server also hosts the Product Data (PD) XML files, which correspond to the elements in the fixture element database.

8.2.2 Client Side

The client side GUI shown in Figure 8.2 has to be downloaded to the user's computer/machine before the system can be run. It comprises of three main portions: (1) Menu class, (2) Viewer class and (3) SIFD module. When it is run, the client will start the application with the Menu class, setting up the GUI and the

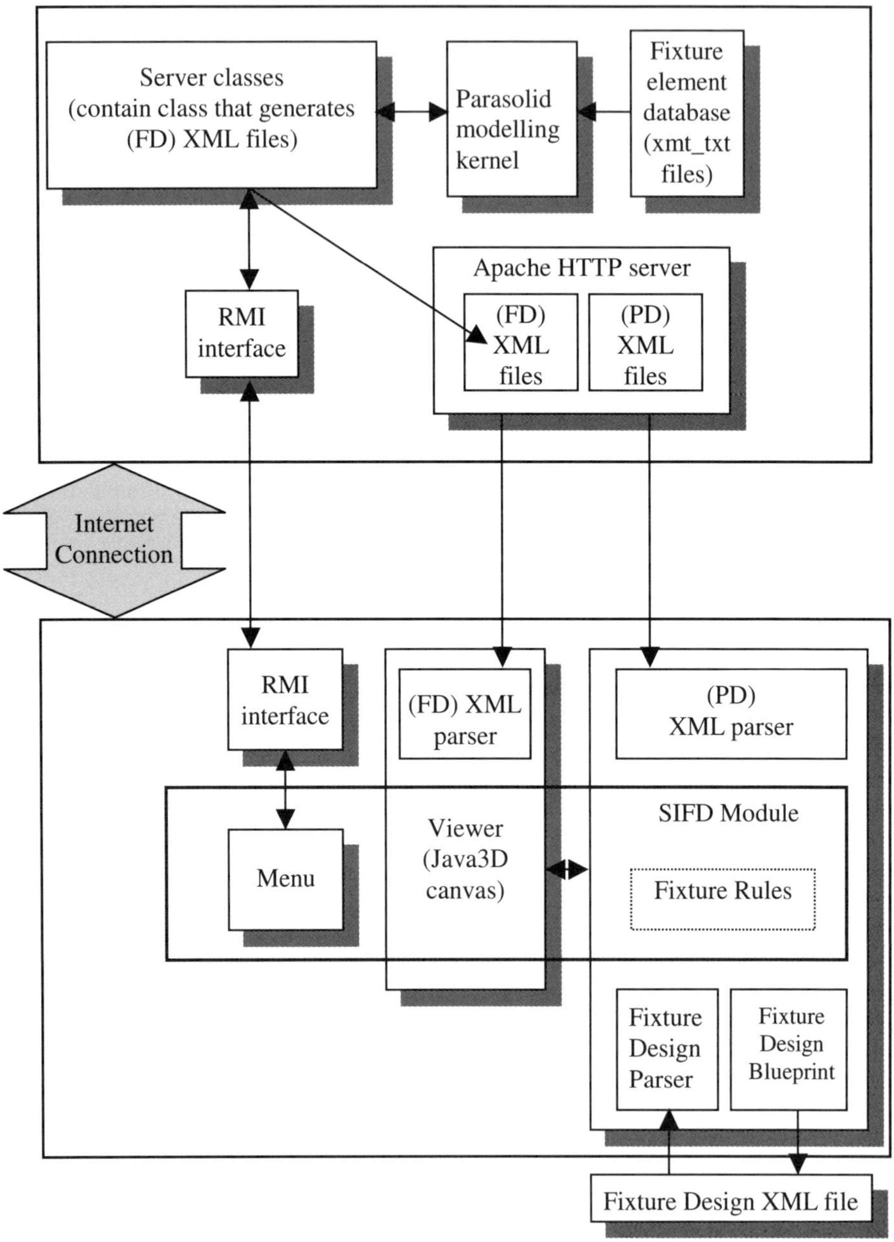

Figure 8.1 Architecture of the complete system

Java3D canvas for modelling, visualization and fixture designing. When a modelling or object selection is made on the menu, a function call is made to the server through the RMI interface.

The Viewer class handles the rendering and visualization of the model design on the Java3D canvas. Through the function calls that are made, Facet Data (FD) XML files are obtained from the server and processed for display on the Java3D canvas.

Upon selection made on the menu, the SIFD module will be activated. The SIFD module can be broken up into five main portions: the main SIFD algorithm, an element information parser, fixture design blueprint class, the fixture design information parser and the different fixture rule algorithms.

The main SIFD algorithm starts the interactive fixture design application and guides the user through the designing process. The interaction with the Viewer class for the visualization of the workpiece and fixture elements on the Java3D canvas is also done here. The element information parser retrieves the Product Data (PD) XML file for the different fixture elements and obtains the information needed by the main SIFD algorithm.

The fixture design blueprint class is responsible for obtaining all the essential data embedded within the completed fixture design. The fixture design information parser retrieves the fixture design XML file that is needed for the design change. The various fixture rule algorithms guide the user through the various stages of the design process and advise him on how to achieve a better solution. This is the portion that gives the application 'intelligence'.

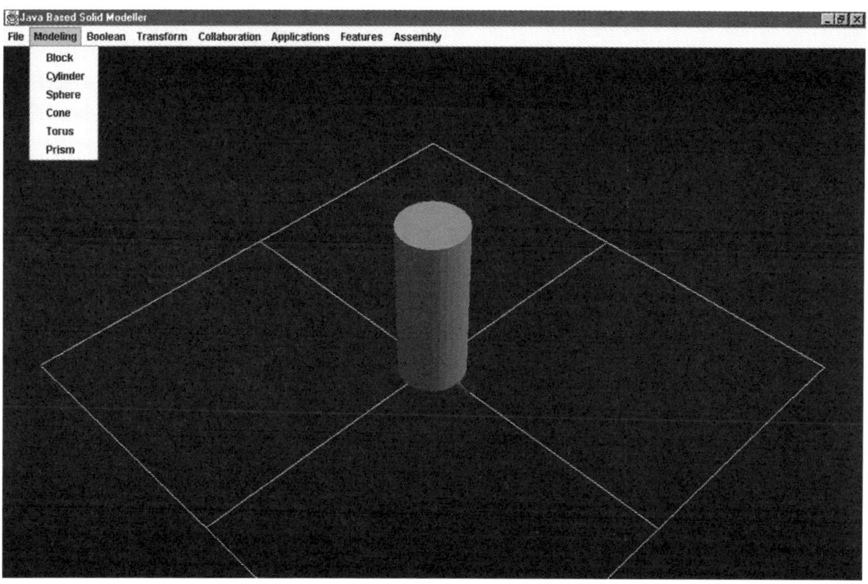

Figure 8.2 The client side Graphical User Interface (GUI)

8.2.3 XML Schemas

Two types of XML schemas currently in the system are used for the file formats of documents that are transferred between the server and the client. Another type of schema is introduced to represent the data detailing the entire fixture design setup. The Document Type Definition (DTD) has to be specified first to enable the XML to be viewed on a Web browser. The DTD describes the data structure of the XML file.

8.2.3.1 Fixture Design XML

The XML file contains all the essential data of a complete fixture design setup. Figure 8.3 shows a basic DTD structure of a fixture design XML file with only a baseplate and a workpiece.

This file contains information about the kind of fixture elements that are present in the system, including the details of the baseplate point that it is attached to, as well as the workpiece faces that are in relation to the element. This XML schema is specific to each element, yet flexible and can be easily expanded to include additional details on the fixture element as well as to include more fixture elements. The XML file is only produced at the end of the fixture design system and placed according to the user's preference.

8.2.3.2 Facet Data XML

This XML file contains facet data, together with some geometric data, needed for visualization and rendering on the Java3D canvas. Figure 8.4 shows the DTD structure of a (FD) XML file.

8.2.3.3 Product Data XML

This XML file contains product data of the elements in the fixture element database. These files are used by the SIFD system during the design process to aid in the selection of elements as well as determination of the display locations of the elements. Additions have been given to some of these files. Figure 8.5 shows the DTD structure of a (PD) XML file.

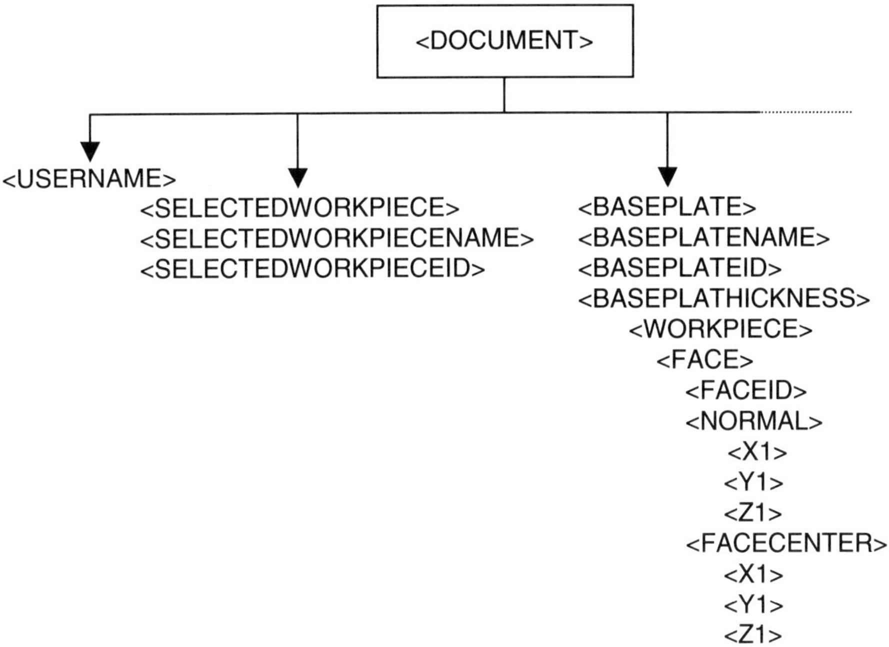

Figure 8.3 DTD of a sample Fixture Design XML file

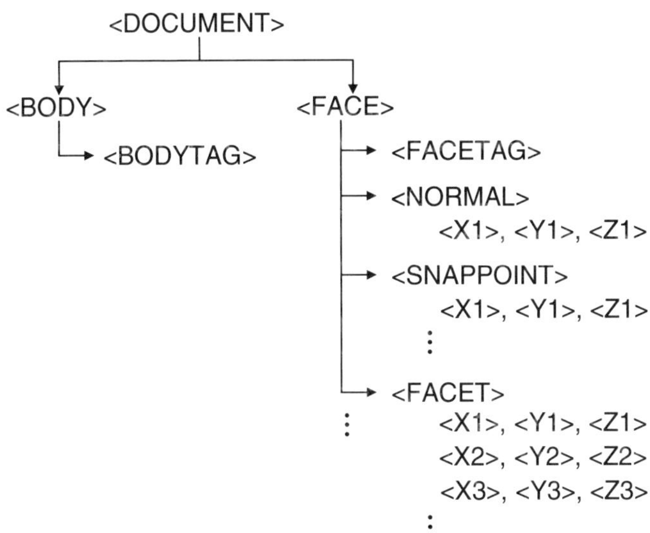

Figure 8.4 DTD of a Facet Data XML file

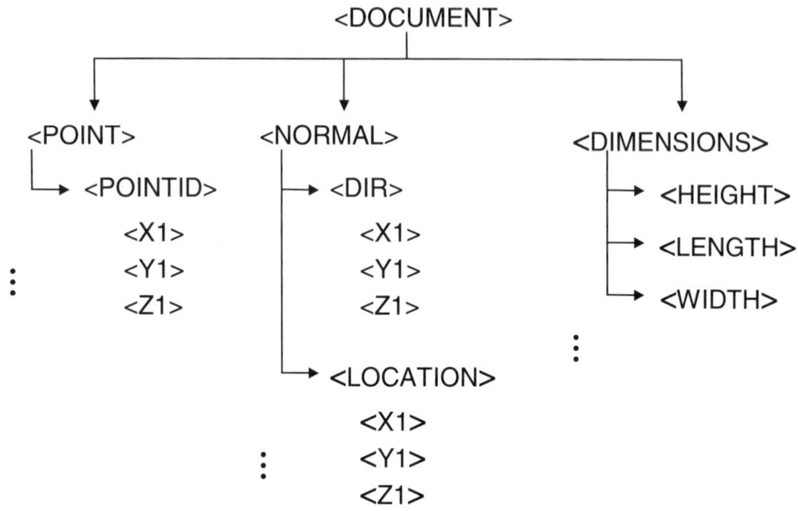

Figure 8.5 DTD of a Product Data XML file

8.2.4 System Application Process

For a clearer and more complete picture of how the entire system architecture is linked together, the whole process of designing a sample fixture system involving only the workpiece and the baseplate will be discussed below. Figure 8.6 shows the steps involved during the whole process.

The RMI interface and the Web server on the server end have to be initialised first before the application can run. The client can be downloaded and run on any remote machine. To begin, the user first has to choose the 'Virtual Fixturing' icon on the menu list.

1. The selection of the workpiece is done on the menu. The information of the workpiece can either be in the form of a Parasolid xmt_txt file (if the workpiece is available from the server's database), or a generated (FD) XML file (if modelling had been done on the modelling interface itself).
2. Once the workpiece has been selected and loaded, the SIFD application starts. The user is asked to choose a baseplate. Once chosen, an element loading function call is initiated on the client. Concurrently, the element information

parser of the SIFD module is invoked to retrieve the (PD) XML file from the Web server. This file is parsed and the information stored for further use.

3. The RMI connection activates the server classes needed to load the baseplate Parasolid file.
4. These classes then utilize the functions on the modelling kernel to load the baseplate xmt_txt file and convert it into triangular facet data. This is done through the use of JNI.
5. The modelling kernel obtains the file of the baseplate from the fixture element database and performs the conversion.
6. The facet data and other geometric information like face normals are arranged into a (FD) XML file on the server and placed on the Web server. This file is then passed to the client side and the initiative is back to the client.
7. The Viewer class accesses the Web server over the Internet and downloads the (FD) XML file that corresponds to the selected baseplate.
8. XML parser within the Viewer class disseminates the facet information and the Java3D scene graph is constructed to represent the baseplate.
9. The baseplate product data obtained previously is associated with the scene graph of the plate so that they can be utilized during fixture design. In this case, the (PD) XML file of the baseplate contains hole information needed for point picking.

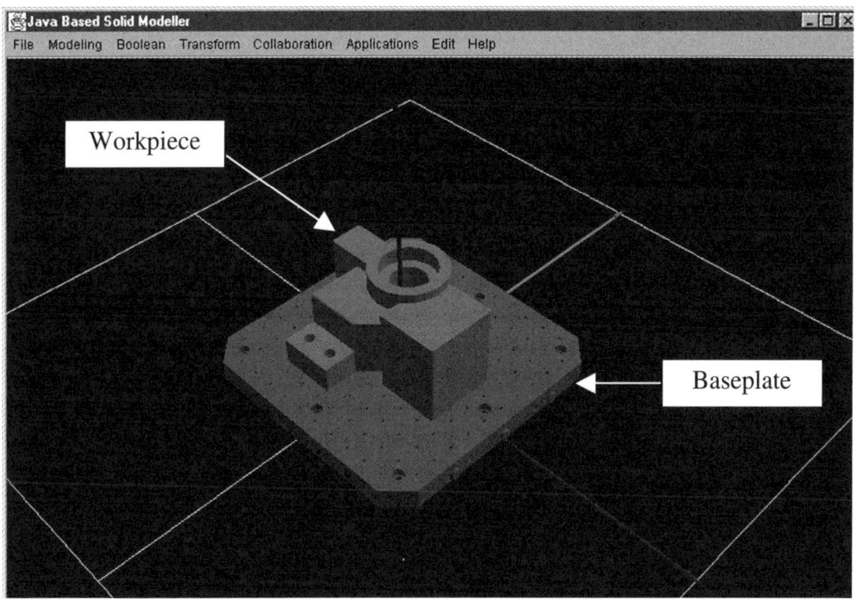

Figure 8.6 Loaded workpiece and baseplate in the Java3D canvas

10. Processing of the scene graph is done and the image rendered in the Java3D canvas (Figure 8.6).
11. Information is gathered and stored during the fixture design process through the use of the fixture design blueprint class. In this case, only the essential information of workpiece and baseplate is stored.
12. Upon confirmation of design, the fixture design XML file is generated.
13. When the fixture design XML file is opened again on another client application, the fixture design parser is invoked. The file is parsed and the relevant (FD) XML files are obtained from the server side and the rendering and visualization steps are repeated again.

8.3 System Structure

The core structure of the IFD module can be summarized in Figure 8.7.

The element selection classes communicate with the main IFD class at different stages of the fixturing process to display graphical dialogues for the user to make element selections. Figure 8.8 shows an example of a graphical dialogue used.

The other portions that constitute the IFD module would include the element database, the (PD) XML files and the element information parser, all of which have been discussed in the previous section.

The IFD module coordinates with the other portions of the client application, like the Viewer class, for the visualization of the fixture design during the whole process. It also contains the algorithm to do the calculations for the riser element selection.

8.3.1 Application Process

The current system is rigid in nature, as the user has to follow a strict sequence of design stages and input of elements is required at each stage. Figure 8.9 gives an overview of the design functions and the sequence of fixturing.

Once the fixture design option is selected, the user has to load a workpiece from the server's database or directly use an existing part created on the modelling interface. Once the workpiece is loaded, a selection dialogue will be displayed to allow the user to select a baseplate. After the baseplate has been selected, the user will have to choose a face on the workpiece to position the top face of the baseplate.

Once the baseplate is loaded, the user selects a support pin from the selection dialogue. After the support pin is selected, the user then chooses a point on the workpiece to position the support pin. The system will then move the workpiece accordingly to accommodate the support pins. Any number of support pins can be added at the discretion of the user.

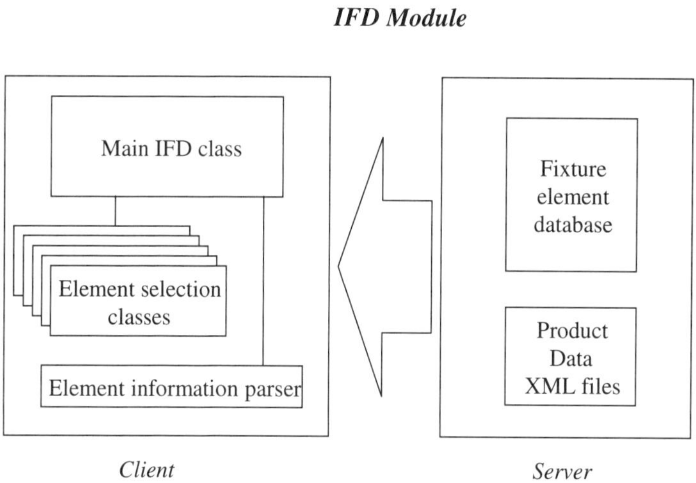

Figure 8.7 Core structure of IFD module

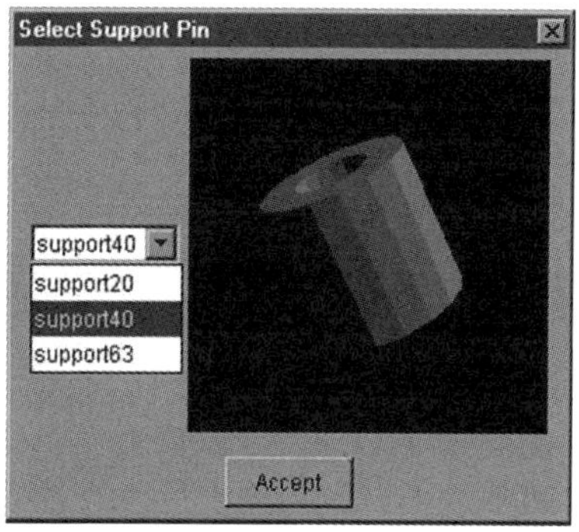

Figure 8.8 A sample element selection dialogue for support pins

Once the support pins are loaded, the selection and placement of locators are done. After the selection of the first locator is done, a vertical face and a point on the baseplate have to be chosen. The workpiece will then be translated accordingly so that the face is in contact with the locator. For the second locator, the user has to select the workpiece face that is parallel to the one used for the first locator. For the third locator, the vertical face chosen has to be perpendicular to the previous two faces. The reasoning behind this locating sequence is so that the 3-2-1 locating principle is upheld. Furthermore, all the locators chosen will have to be taller than the support pins or else the system will not allow the user to proceed.

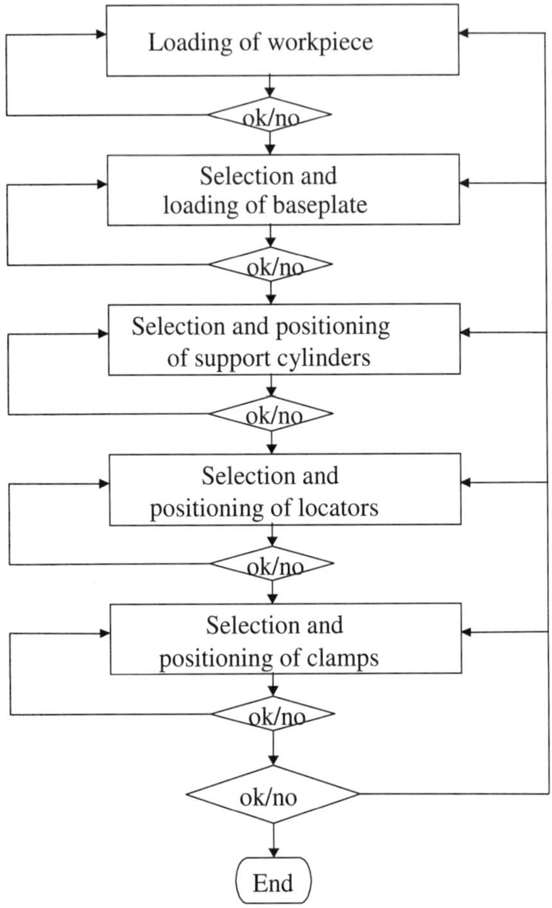

Figure 8.9 The design sequence of the IFD system

Finally, the selection and placement of clamps are done. There are two types of clamps currently available in the system: hook clamps and side clamps. For hook clamps, the user has to select a horizontal face for clamping, a vertical face to orientate the clamp and a baseplate point to place the clamp. Using the horizontal face height as a reference, the type and number of risers needed to elevate the clamp is obtained. For side clamps, the user only has to select a vertical workpiece face for clamping. The system uses side clamps for faces that are perpendicular to the gridlines of the baseplate, for faces that are offset at an angle to the gridlines, an adjustable stop is recommended.

As seen from Figure 8.9, there is an undo option after each selection is made. Furthermore, the user also has an option to undo the selections he has made at the end of the fixture design process and return to the beginning of any one of the five design stages.

The current system is also able to do simple analysis on the points selected for placement. If the distance perpendicular to the face is too near or too far from the workpiece, the system is able to identify the problem and notify the user accordingly.

8.3.2 Limitations

As a prototype system, the fixture design module serves only to illustrate the feasibility of implementing a fixture design system on the Internet based on the client-server architecture. It is a very basic system and thus has quite a substantial number of design limitations.

First of all, the fixture design module does not have any fixture design saving and retrieval capabilities. This problem is significant as there is no structure in place to capture all the important features of the fixture design system. Once the client application is closed, the design is lost.

The system also has not looked into the possibility of fixture design change and thus does not make any accommodations in this area. Currently, the user can only build a fixture design from scratch and not able to make deletions and amendments to existing ones. In terms of achieving a truly collaborative design environment, this is one area that needs to be looked into.

Currently, the system is not user-friendly. There are no guidelines for a user with little or no knowledge of fixture design to follow when he is performing fixture design.

The system is only able to handle prismatic parts due to insufficient information provided by the present (FD) XML files. Due to the heavy reliance on face normal and face centres for many of the functions applied, workpieces with curved surfaces will not work properly on the present system.

Lastly, the algorithm needs to be improved to handle more complex workpieces. Currently, the system only works with support pins that have the same height. This means that the pins can only prop up uniformly flat bottom surfaces properly. Locators and clamps need vertical faces and horizontal faces for positioning, which will rule out slanted surfaces altogether.

8.3.3 Storage Structure

The storage structure is implemented through the use of a few classes. These are illustrated in Figure 8.10.

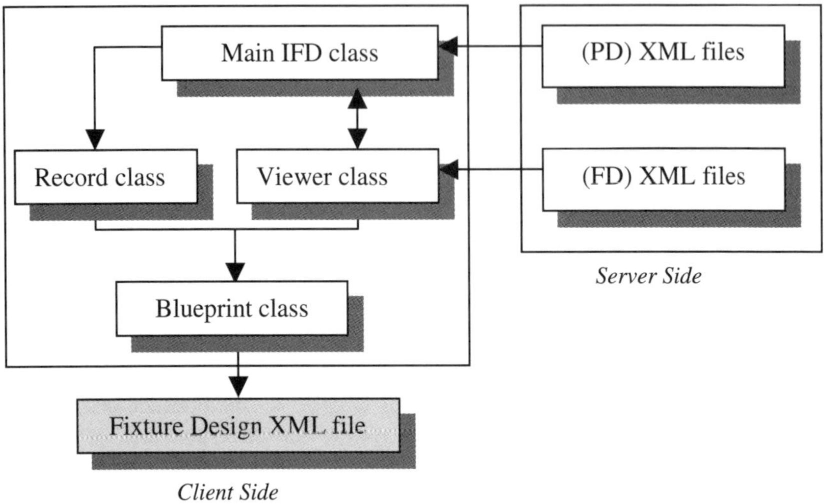

Figure 8.10 Structure of storage system

Two classes are used to extract the essential information from the main IFD class during the fixture design process. The Viewer class extracts element information from the (FD) XML files, after which only the relevant information like body tags, face tags and face normals are filtered by the Main IFD class and stored in the Blueprint class. The Record class mainly extracts the relevant information obtained from the (PD) XML file and stores it in the Blueprint class. Upon confirmation of the fixture design, the user is prompted to name the directory and the file used for saving. Figure 8.11 shows the file saving dialogue box.

Once this save command is executed, the Blueprint class will take all the data and format it into a Fixture Design XML file. This file is compact and can be easily stored on a diskette or transferred through the Internet. With this capability, interaction between programmers is greatly enhanced.

8.3.4 Retrieval Structure

The classes involved in the whole retrieval structure are shown in Figure 8.12. The Fixture Assembly class is activated by the Menu class when the user selects the 'Retrieve Fixture Design' option. This class then utilizes the Fixture Design XML file to find the correct (FD) XML files and (PD) XML files on the server needed for visualization purposes. The Viewer class handles the visualization while the Object Record class stores the number of objects currently present on the Java3D canvas. This is needed for deleting and changing fixture design, which will be discussed in the later sections.

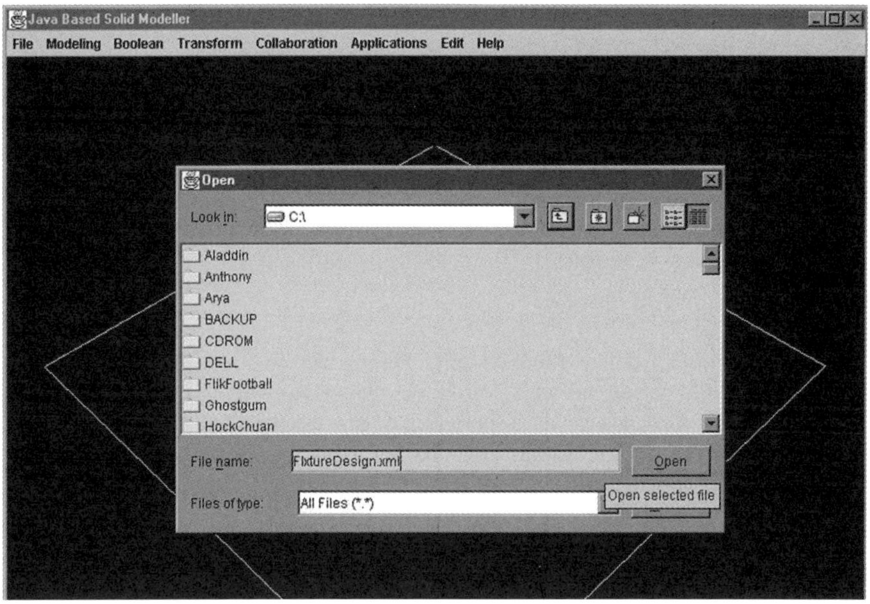

Figure 8.11 Dialogue box used for file saving

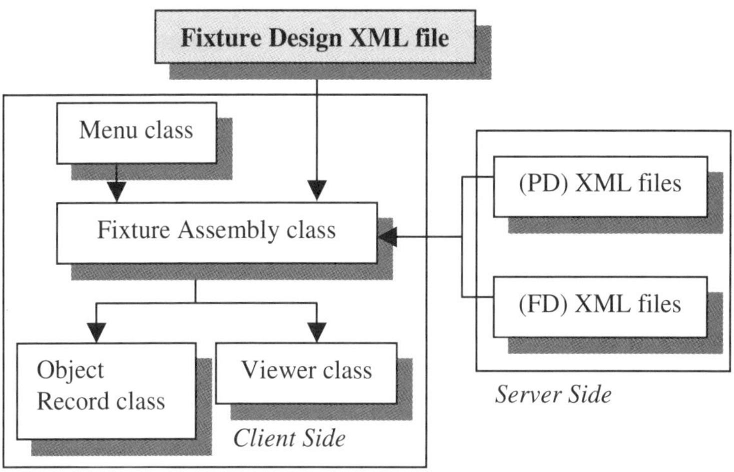

Figure 8.12 Structure of Retrieval System

8.3.5 Changing a Design

In order to bring fixture designers across the globe to perform fixturing activities in real time, the basic infrastructure for file sharing and transfer, as well as fixture design change capabilities must be implemented. These developments also mark the first step in developing a truly collaborative design environment. This section looks into the object deleting function and how this is modified and coordinated with the Fixture Design XML file to achieve fixture design change capabilities.

8.3.5.1 Object Delete Function

The previous object delete function only allows the user to remove the loaded elements from the display, as well as from the Object Record class, at the end of the fixture design process. However, a limitation of this function is that it only removes the deleted object from the display area and not from the Fixture Design XML file. Hence, retrieved files will still include the 'deleted' object in the setup.

The new system only allows the various undo functions to be used for the editing of elements. Once these functions are invoked, the user can choose to restart at any one of the fixture design stages. Such a process ensures that the Fixture Design XML file is constantly updated on any changes made to the previous design.

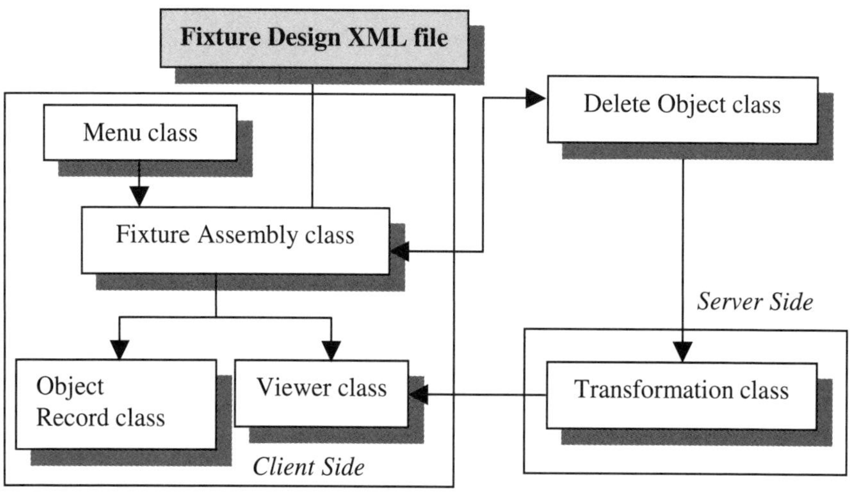

Figure 8.13 Structure of Deleting System

The Object Delete Function is now modified to remove objects only from retrieved fixture design files. Once an object has been selected for deletion, the system will utilize the Fixture Assembly class to determine the type of element object. After the object is identified, the system will prompt the user to delete all of that particular element type. The object is removed from display through the Viewer class and also removed from the Object Record class. If the element is a support pin or locator, the system will utilize the transformation class to move the workpiece to its original position.

Figure 8.13 shows the classes involving in deleting. Figure 8.14 shows the deleting process for support pins, locators and stops. Figure 8.15 shows the deleting process for side clamps and hook clamps.

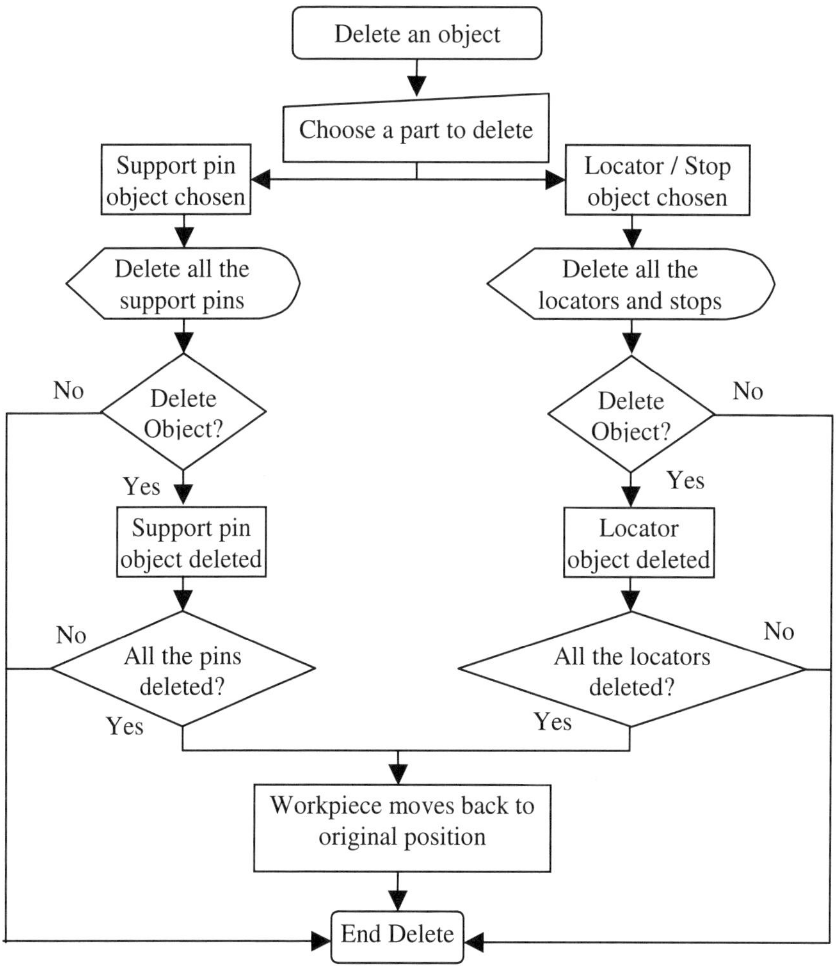

Figure 8.14 Deleting process for support pins, locators and stops

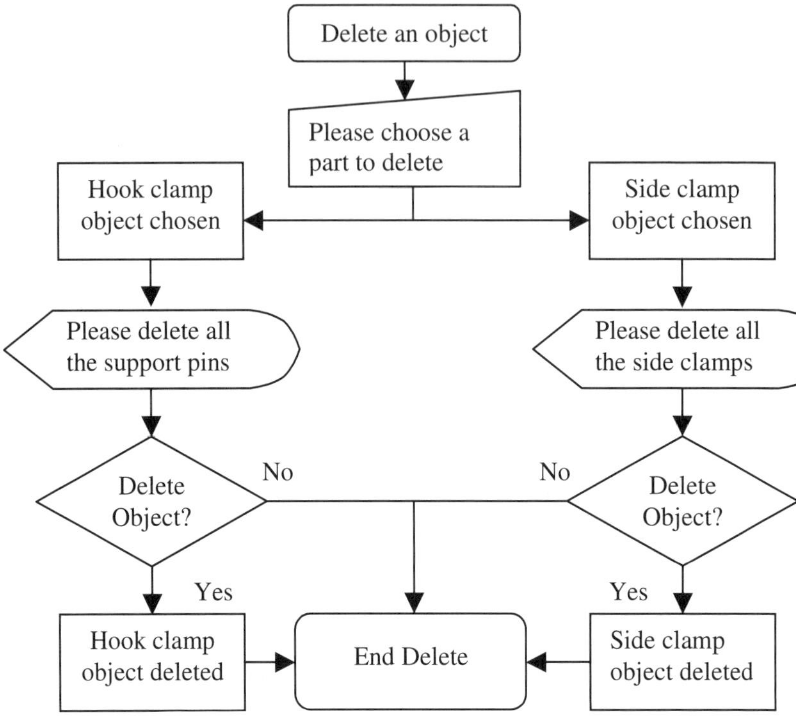

Figure 8.15 Deleting process for hook clamps and side clamps

8.3.5.2 Design Process for Existing System

A workpiece has to go through the entire procedure illustrated under the application process of the SIFD system. However, for an existing fixture design system, the user can perform design change to the original fixture solution only if all the elements of a particular stage have been deleted (e.g., if all the support pins or all the side clamps are deleted, a user can add support pins). The user only has to go through that design stage to replace the deleted elements. Figure 8.16 shows the design process.

To coordinate the information stored in the original fixture design system and the new information created, a Parse-Existing-System class is created. This class extracts and stores the information from the original fixture solution. After changes have been made, the new data is stored in the Blueprint class. The Blueprint class will gather the information of the non-deleted elements of the original solution from the Parse-Existing-System class and produce the new Fixture Design XML file. Figure 8.17 shows the saving process.

The rationale for installing such a rigid sequence for deleting and system designing is such that the Fixture Design XML file will always be complete. This means that the basic structure of a fixture design: baseplate, support pins, locators and clamps will be intact despite whatever modifications that are done to the design.

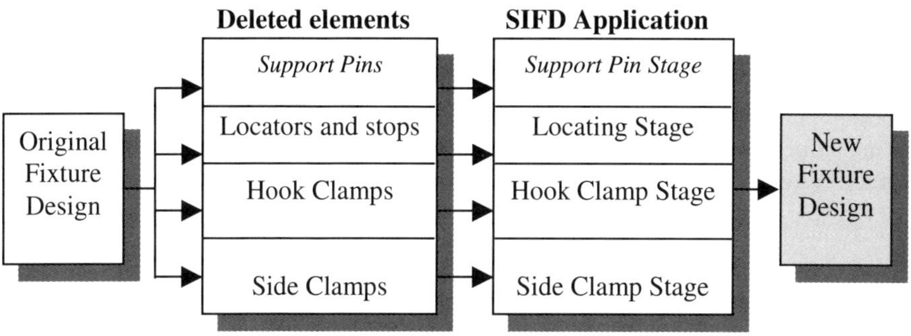

Figure 8.16 Process for design change of original fixture solution

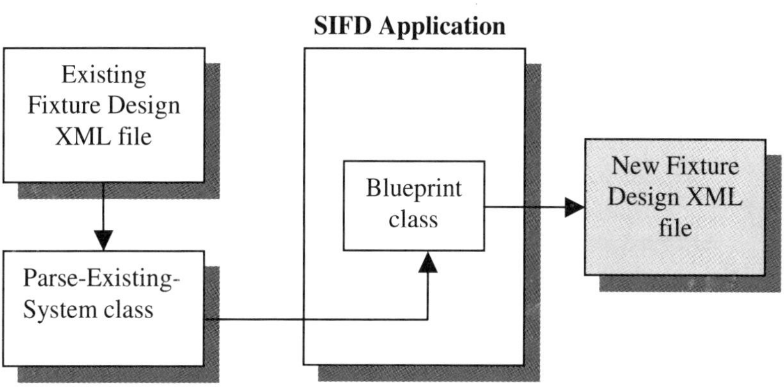

Data from non-deleted elements

Figure 8.17 Storage process system

8.4 System Architecture

The architecture of the rule-based system uses the SIFD main application as the main spine along which essential data is extracted. This data is obtained from the application as the user performs the fixture design and is stored as XML files. Using the data from these files, fixture rules are implemented and analysis that is done is fed back into the main application. According to the analysis done, the system will advise the user accordingly as to which course of action should be undertaken.

This structure forms the platform on which more rules and data can be added with minimum disruption to the SIFD main application. In addition, repetition of coding can be cut down, as the rules implemented on the same element will be using information from the same XML files. Figure 8.18 describes the repetitive architecture of a rule-based fixture design system.

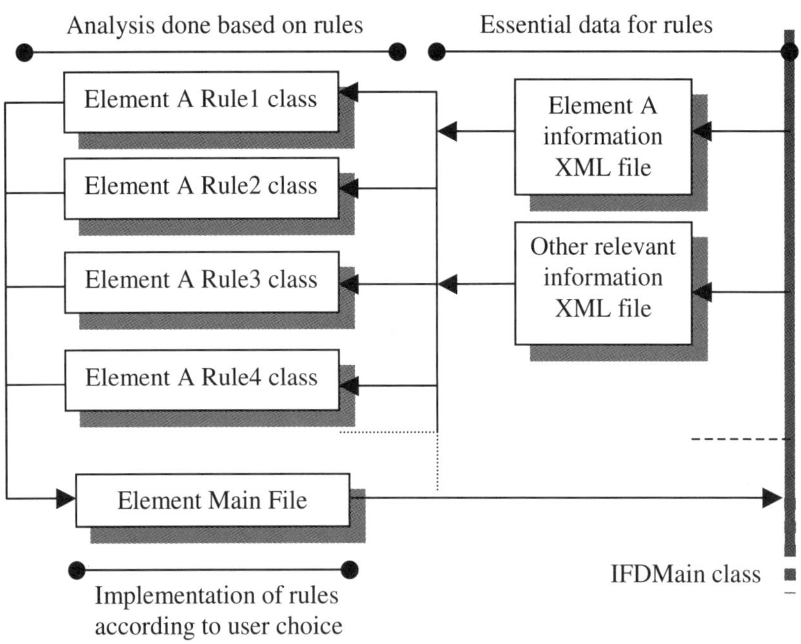

Figure 8.18 Architecture of a rule-based system

8.4.1 Support Pin Rules

One general rule for placing support pins is that the triangular area formed by the three supports is the largest possible for the face selected. This is to ensure greater stability of the workpiece. This is illustrated in Figure 8.19.

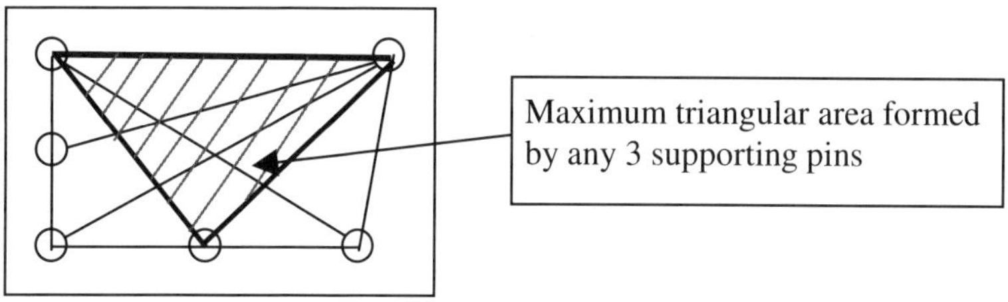

Figure 8.19 Top view of workpiece

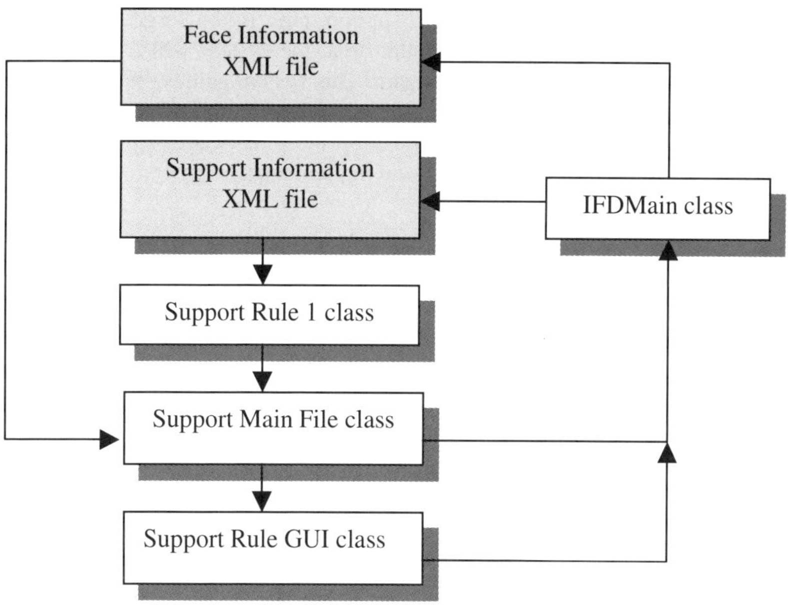

Figure 8.20 Support Rule Implementation structure

The support pin rule structure basically follows the general architecture for rule implementation mentioned above. The Support Information XML file contains the position coordinates of the support pins and the Face Information XML file contains the triangular face area. The Support Rule1 class uses the data from both files to perform the analysis.

The system calculates the triangular area formed by the three pins and compares with the maximum triangular area formed by the supporting face. Once the analysis is done, the results are sent to the Support Rule GUI class, which then displays the result on the graphical user interface. At the same time, the Support Main File class interacts with the IFDMain class, guiding the user towards better design implementation. Figure 8.20 gives a closer look at the classes involved.

8.4.1.1 Support Rule Implementation Process

The rule is activated only after the third support pin is selected. Upon confirmation of selection, calculation is done to determine the triangular area formed by the three support pins. This is compared with the maximum triangular area that can be formed from the workpiece face. If the area is within an acceptable range of values, the system will issue a statement saying that the support pins cover an acceptable area of the face.

If the triangular area formed by the support pins is too large or too small compared with the maximum face triangular area, the system will ask if the user would like to reposition the support pins again. This system guides the user towards building a better fixture design but is flexible enough to let the user make the final decision. The user can then choose to reposition the pins again or proceed to the next stage. Figure 8.21 shows the support rule implementation process.

8.4.2 Locator Rules

The rule implemented here ensures that the two locating pins are positioned as far apart as possible on the major locating face. The major locating face is defined as the face that takes the cutting load and the minor locating face is, in general, perpendicular to the major locating and supporting face. The rationale for keeping the pins far apart is so that the resultant force vector (resultant of the cutting forces) is contained within the two pins.

As both the first locator and the second locator are placed on the face with the same outward normal, the system automatically assumes that the two locators are placed on the major locating face. Figure 8.22 shows the major and minor locating faces of the workpiece.

The locator rules structure follows the support rules structure with one notable exception: there is an extra Face Length class to calculate the maximum horizontal length of the major locating face. This change is reflected in Figure 8.23.

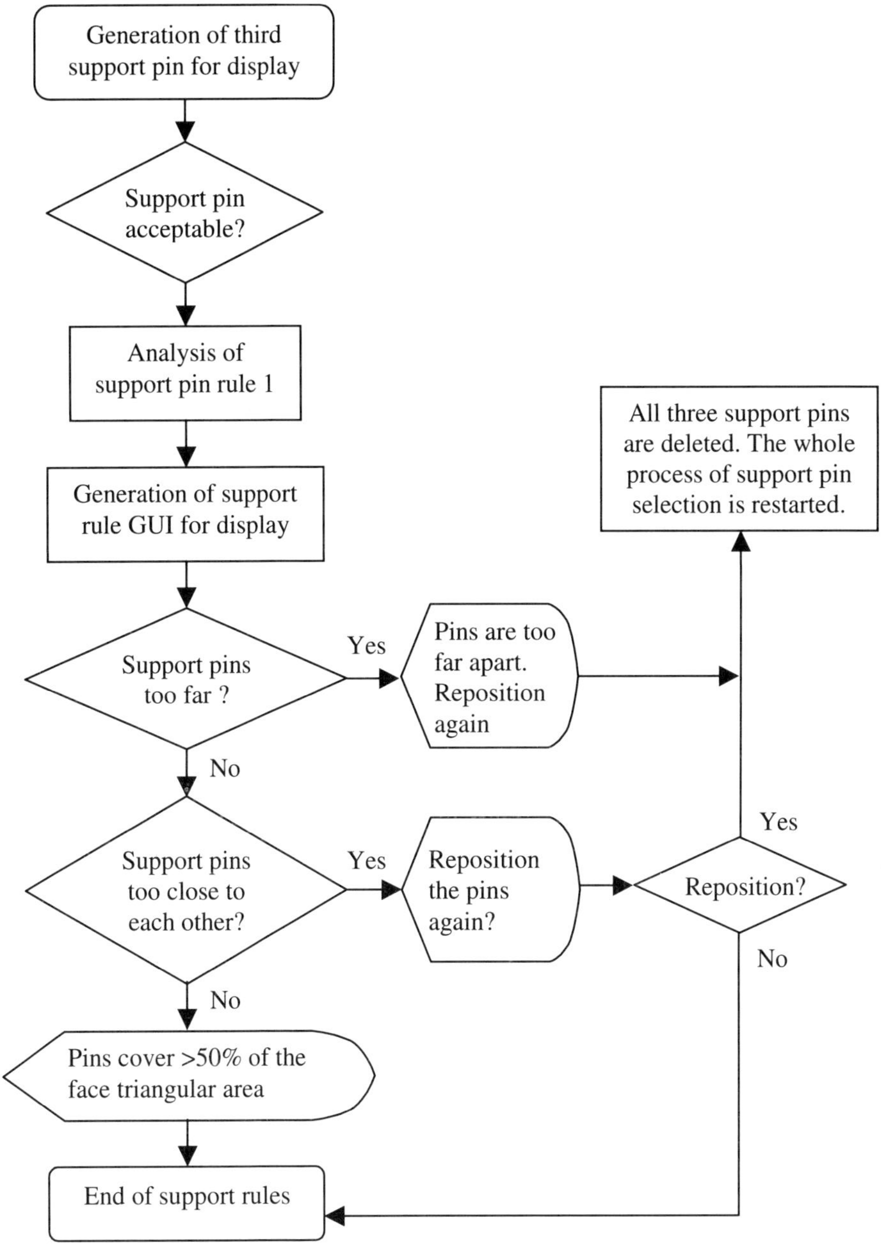

Figure 8.21 Support Rule 1 implementation process

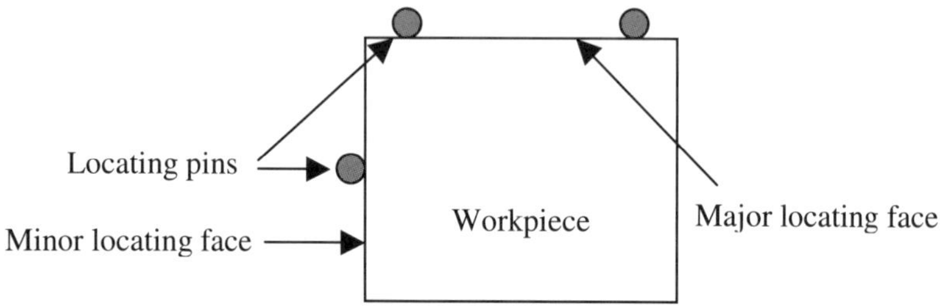

Figure 8.22 Top view of major and minor locating faces

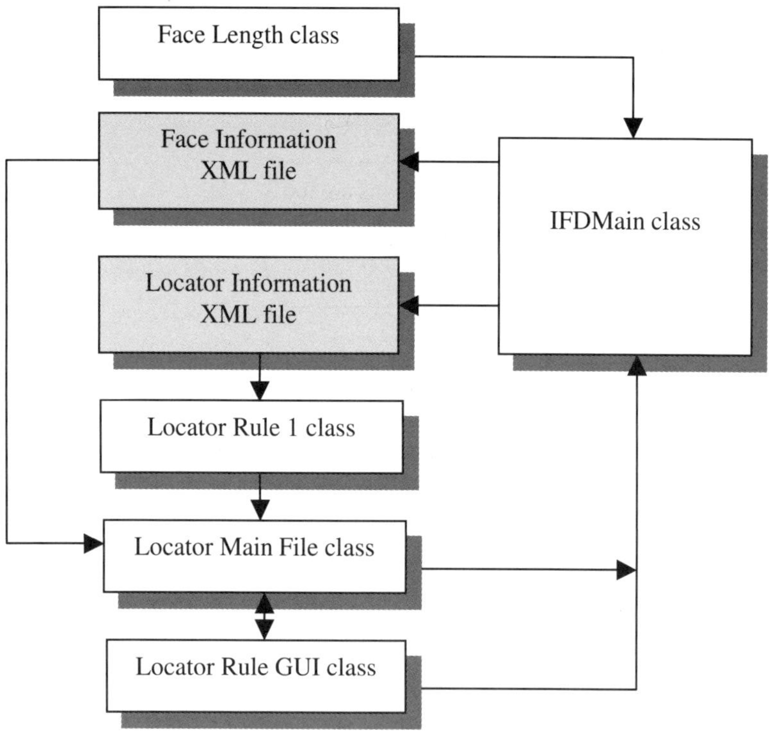

Figure 8.23 Locator Rule implementation structure

8.4.2.1 Locator Rule Implementation Process

The rule follows closely with the process of locating the workpiece. After the selection and positioning of the second locator is confirmed, calculation is performed to determine the distance between the two locators. This is compared with the maximum horizontal face length of the major locating face.

If the distance between the two locators is greater than the face length, or if the locators are not placed on the face, the system will issue a warning message to notify the user of the situation.

If the face length is greater than the distance between the two locators, the system will inform the user that the locators can be placed further towards the edge of the face. The user can thus make the amendment, provided that there are locating holes available. The rule can still be implemented if the two locators are situated at different locating faces. The final decision on the repositioning of the locators is left to the user, as the aim of the system is to guide, not to direct or enforce. Figure 8.24 shows the locator rule implementation process.

8.4.3 Clamp Rules

The clamping face is selected such that when it is clamped, the clamp can hold the workpiece rigidly to counteract all the cutting forces. Thus clamping faces, in general, will be opposite to the locating faces. Furthermore, in addition to this rule, it is stated that clamping forces should be directed within an area contained by the locating pins as shown in Figure 8.25.

The system aims to implement these requirements with the construction of two clamping rules. Clamp Rule 1: Preferable for edge clamps to be placed opposite to the major locating face. Clamp Rule 2: Clamping forces should be directed to within an area contained by the locating pins. A check system is in place and the system will only allow the implementation of the second rule if the first rule is not violated and will strongly recommend the user to implement the first rule together with the second rule.

Clamp Rule 1 is completely embedded within the IFDMain class as it utilizes and compares the face normals that are already available in the class. The Clamp Rule 2 structure generally follows the support rule structure except that the data is obtained from the clamp information XML file and the locator information file. Figure 8.26 shows the Clamp Rule 2 structure.

8.4.3.1 Side Clamp Rule Implementation Process

The first rule is implemented by checking the outward normals of the major locating face and the clamping face. If the two normals are opposite to each other, the rule is satisfied. If not, the system will ask if the user would like to choose another face.

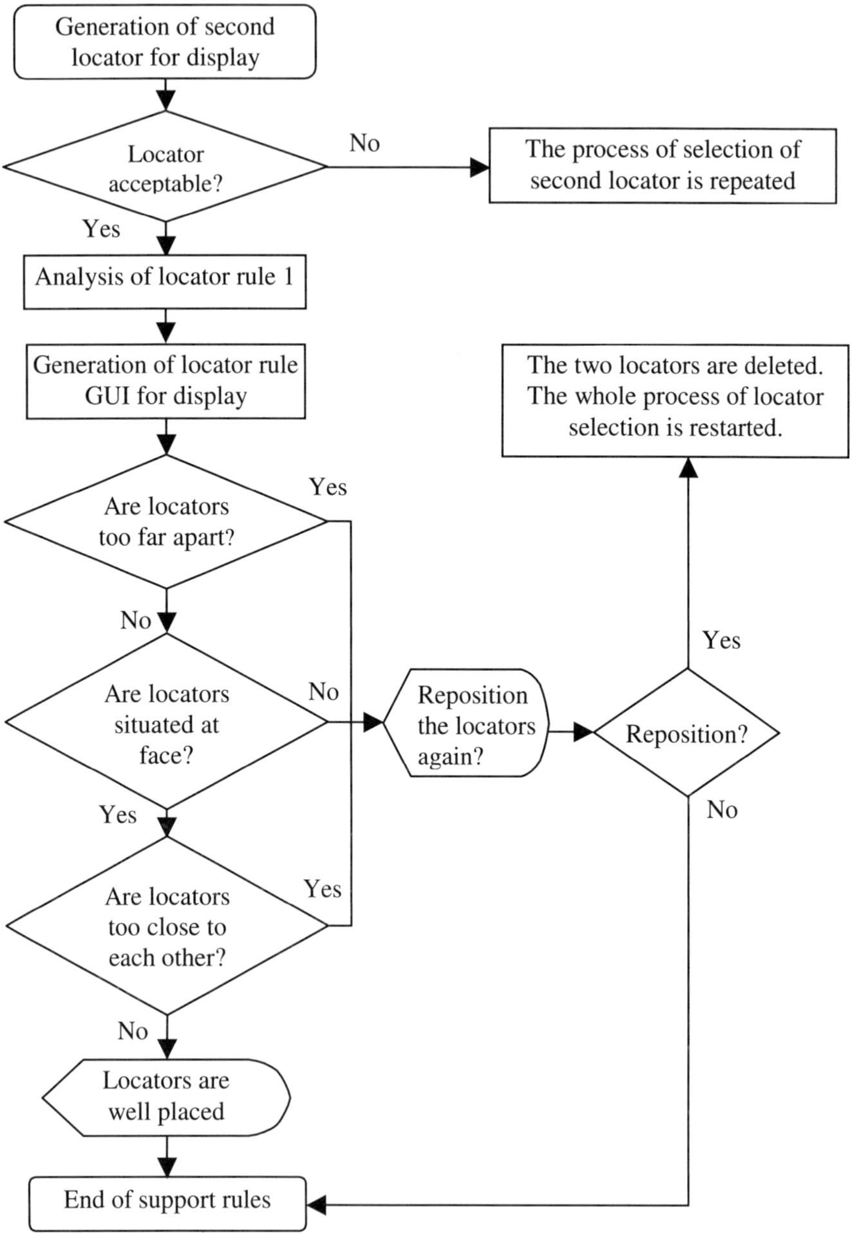

Figure 8.24 Locator Rule 1 implementation process

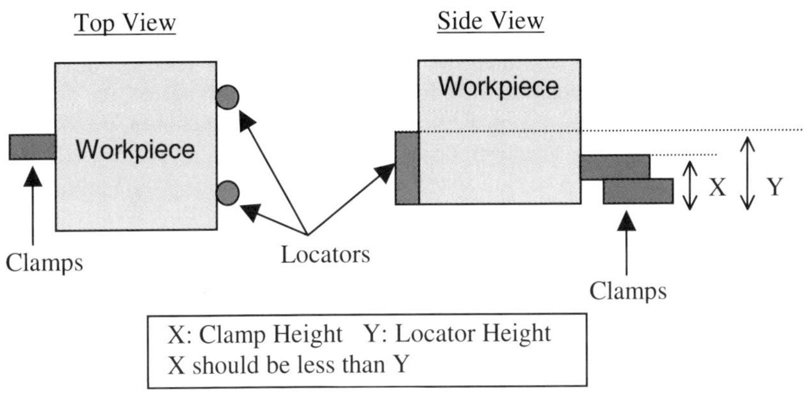

Figure 8.25 Edge clamps placed opposite to the major locating face and within area contained by locating pins

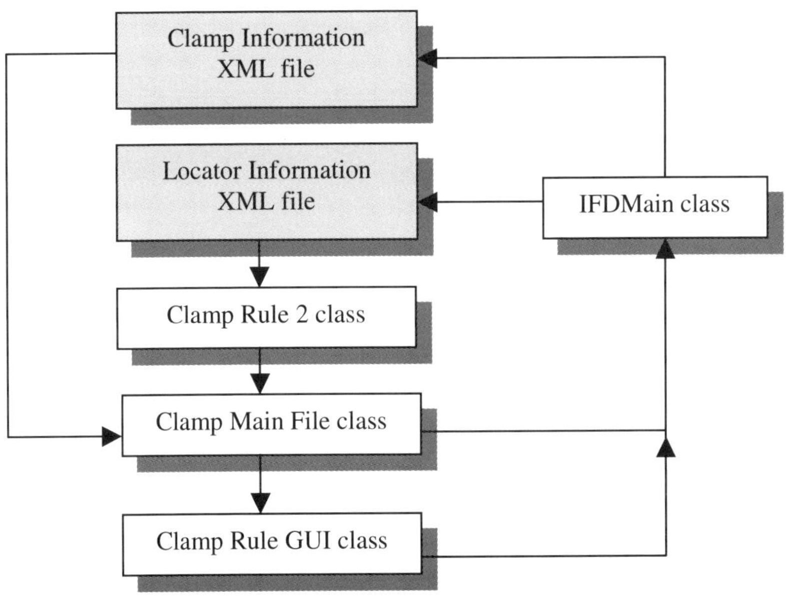

Figure 8.26 Clamp Rule 2 implementation structure

The second rule first checks if the first one is satisfied. If yes, it will check if the clamp is contained within the locating pins area. If the clamp is not within the area, the system will issue a warning and ask the user to reposition the side clamp.

If the locator height is less than the clamp height, the system will advise the user to replace the original locators with longer ones. If the user agrees, the clamp is used and the locating process is repeated. After the locators are chosen, the user can either perform the clamp selection process again or skip it, which will then complete the whole design. The flowcharts showing the implementation process are presented in Figures 8.27 and 8.28.

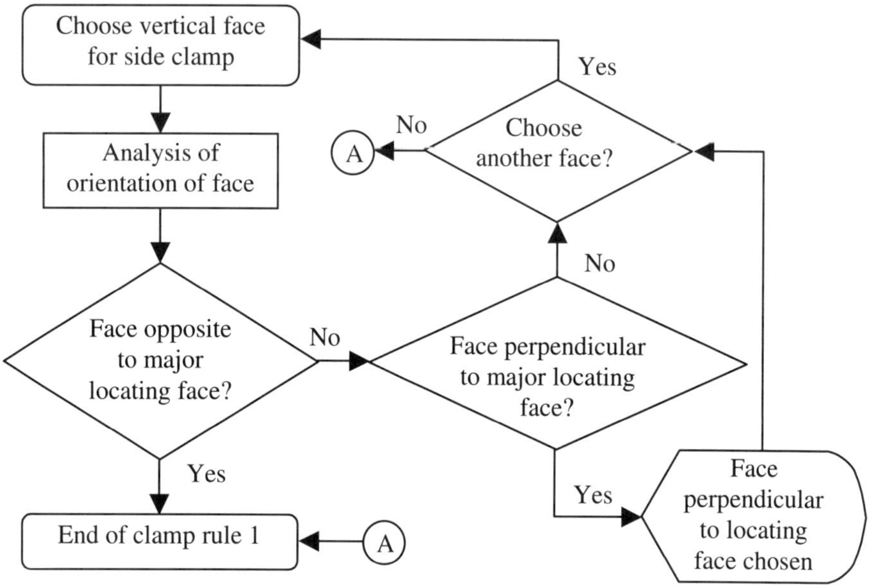

Figure 8.27 Clamp Rule 1 implementation process

8.4.4 Limitations of SIFD Module

It must be stressed that the rules implemented in the SIFD module only serve to illustrate the feasibility of offering rule-based functionalities and lay the groundwork for future rule implementations and improvements. It is a basic system and possesses a number of design limitations.

The support rules can be implemented only if the area of the surface is known and currently the information is manually added into the XML file. Furthermore, the system is not able to detect if the support pin is not within the confines of the supporting plane surface. Future developments could include research on the kernel functionalities in order to derive the face area and collision detection to check if the support pin is supporting the surfaces.

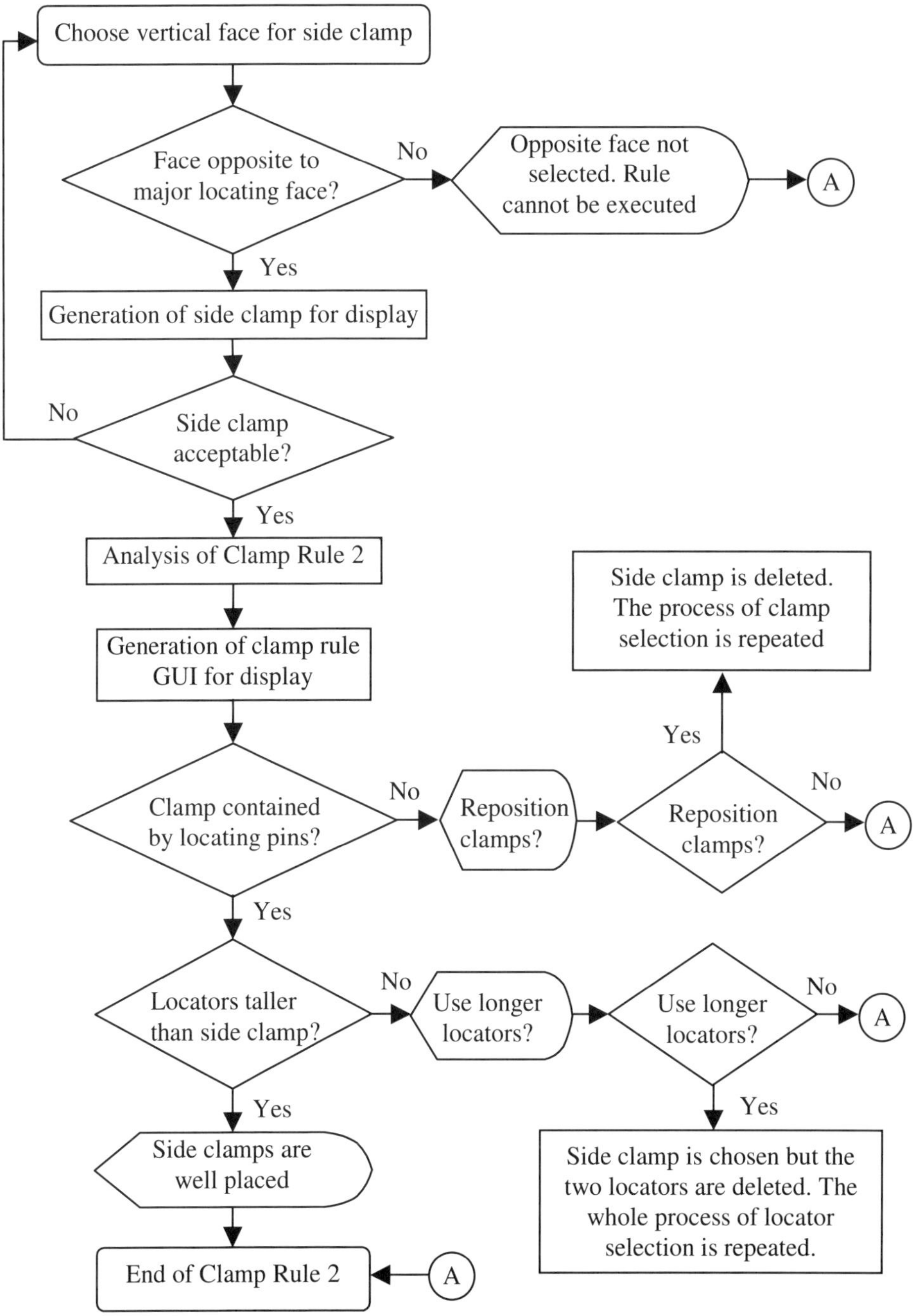

Figure 8.28 Clamp Rule 2 Implementation Process

The locator rule can be implemented on any polygonal planar surface as it relies heavily on the snap points. Therefore the rule cannot work if the surface has curved edges with no snap points.

The first clamp rule is able to handle any planar surface with a defined normal and is hence not workable with curved surfaces without defined normals.

Most of the rules do not depend solely on information regarding the workpiece and fixture elements. More detailed information on other constraints like cutting forces and tolerance must be included before these rules can be implemented.

8.5 An Illustrative Case Study

The workpiece used in this case study is shown in Figure 8.29. This workpiece has a flat planar bottom and polygonal surfaces, satisfying requirements needed by the fixture rules. The curved and non-polygonal surfaces are not used during the fixture designing. All the four rules mentioned in the previous sections will be illustrated here.

Figure 8.29 Workpiece sample

Once the SIFD module is invoked, a dialogue box will appear requesting the user to choose between: (1) Loading a workpiece from the server's database, (2) Using an existing model on the modelling interface or (3) Using a previous fixture design setup uploaded onto the modelling interface. An SIFD status frame is displayed to show the designing stage the user is in. All these modifications are reflected in Figure 8.30. The choice made here is to load a workpiece. Figure 8.31 shows the workpiece being loaded.

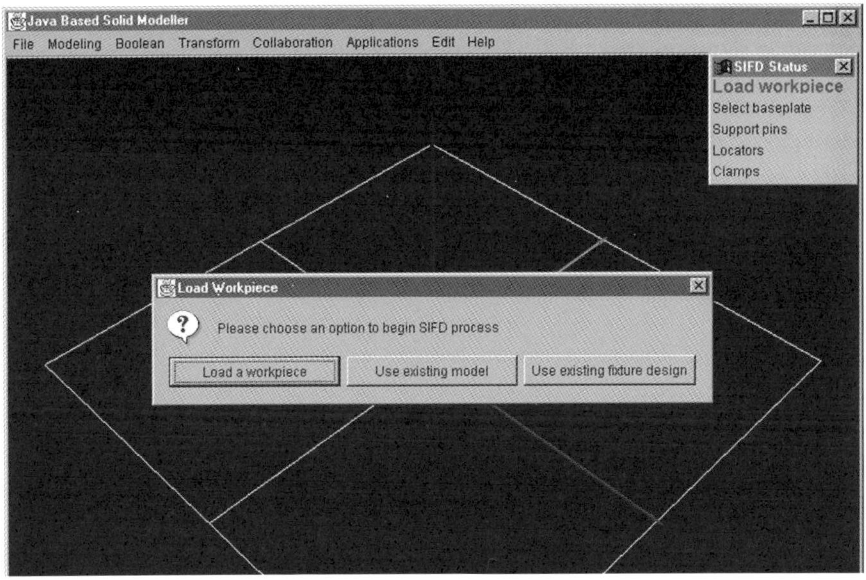

Figure 8.30 The initialisation of SIFD process

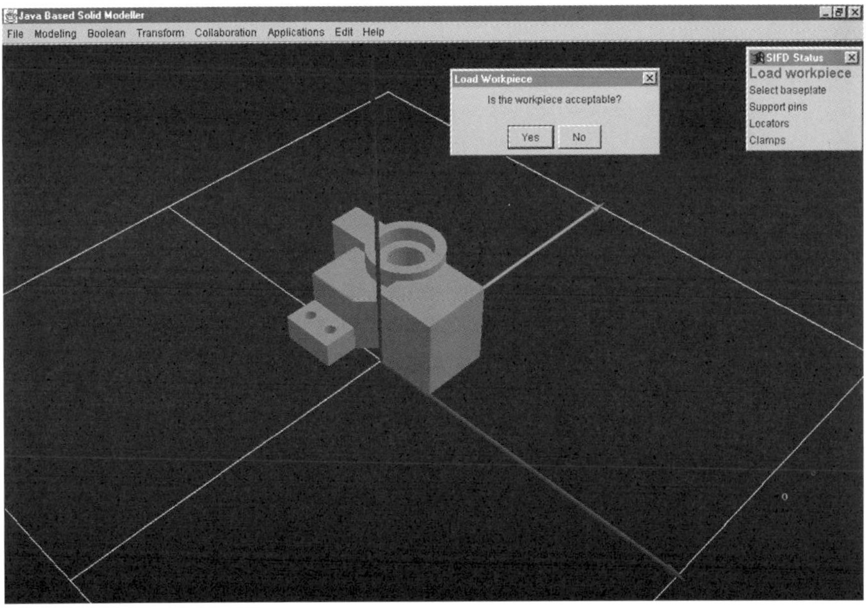

Figure 8.31 Loaded workpiece on Java3D canvas

Once the workpiece has been loaded, a fixture rules selection dialogue will appear as shown is Figure 8.32. The rules available in the system are shown on the left window and the rules that the user has selected for use will be shown on the right window. The bottom window gives the definition of the rules when the definition button is hit.

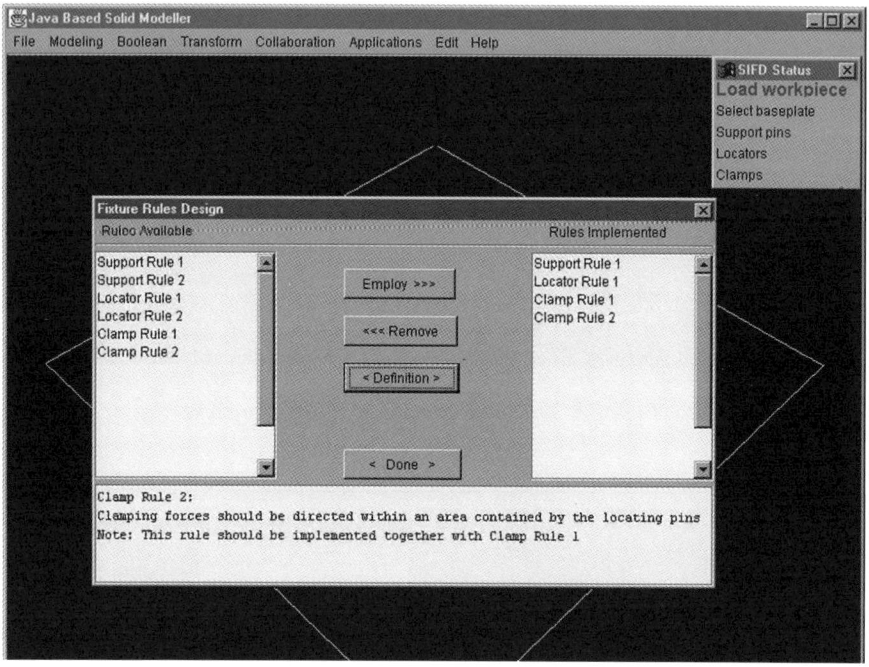

Figure 8.32 Fixture rules selection dialogue

Once the rules are selected, an element selection dialogue will appear for a baseplate selection. A rule status frame is also displayed to reflect the rules that are selected and the rule that is currently being used in the design process will be highlighted (Figure 8.33).

After the baseplate has been loaded and accepted, support pin selection is made. Upon selection and confirmation of the third pin, the first support rule is activated (Figure 8.34). The system will display the information used in the analysis. In this case, the system detects that the pins are placed too far apart and tells the user to reposition the pins again (Figure 8.35). If the area of the support pins is within an acceptable range of values (Figure 8.36), the system will carry on with the design process (Figure 8.37).

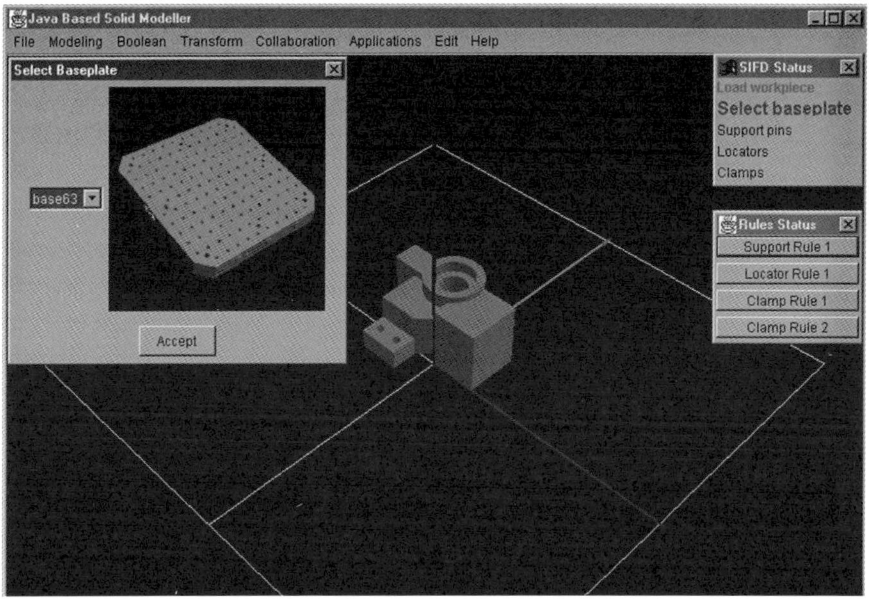

Figure 8.33 Baseplate selection dialogue

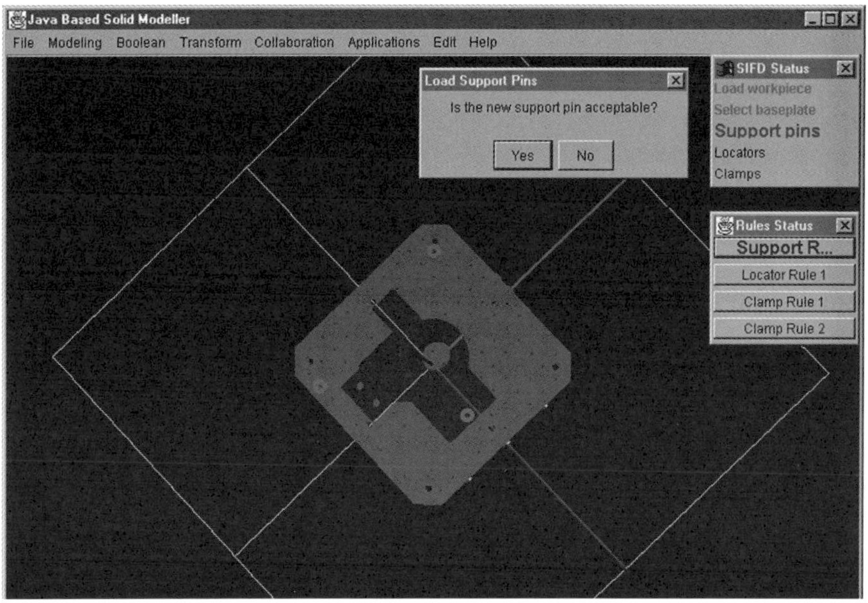

Figure 8.34 Selected support cylinders loaded and positioned (unacceptable)

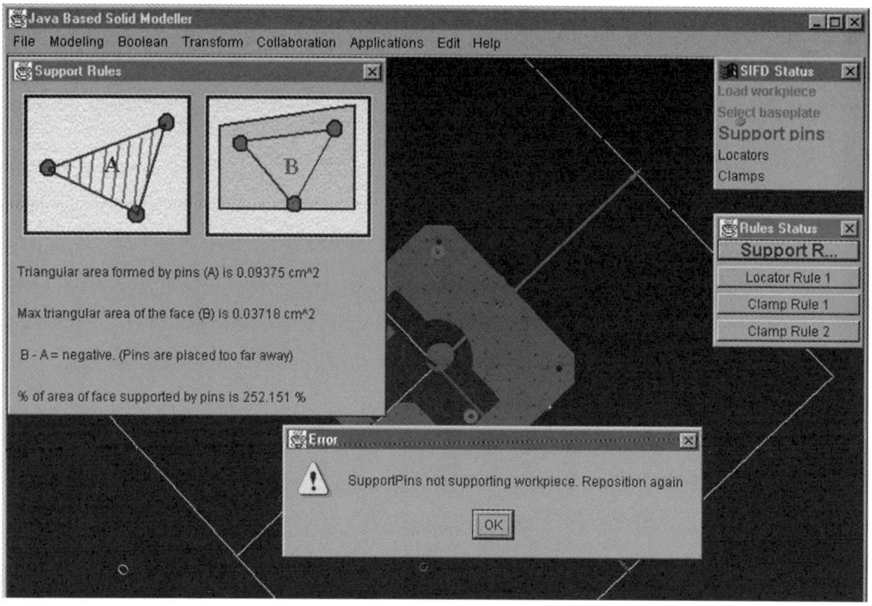

Figure 8.35 Analysis done on support pins based on Support Rule 1

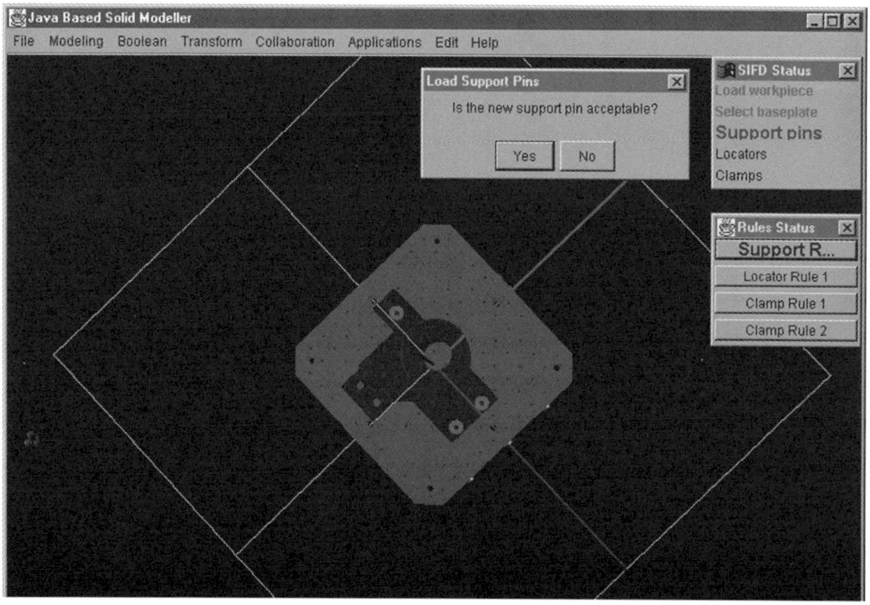

Figure 8.36 New support pin position (acceptable)

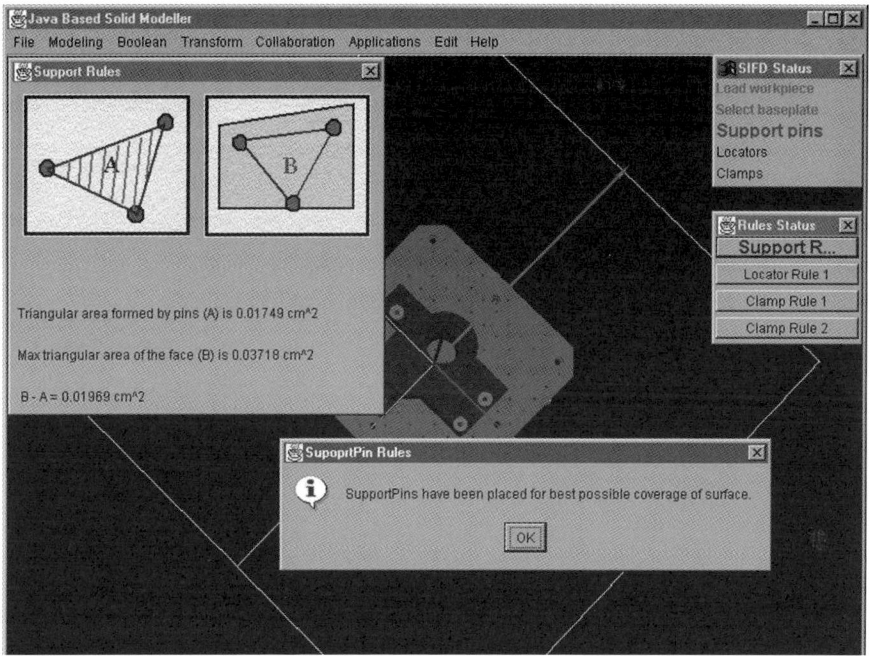

Figure 8.37 Analysis done on new acceptable support pin positions

The locator selection process is shown in the next few figures. Figure 8.38 shows the original locating positions and Figure 8.39 shows the locator rule implementation process. The system detects that the locators can be placed further apart for better location and asks if the user wants to reposition the locators again. Figure 8.40 shows the response that is given when the locators are repositioned and the locator placements are in accordance with the rule.

Rules are activated when the side clamp selection process is done (Figure 8.41). The next few figures will illustrate the process that is guided by the two clamp rules. Figure 8.42 shows Clamp Rule 1 is invoked when the user chooses a face (shown in red) that is perpendicular to the major locating face. The user is then advised to choose another face opposite to the locating face. Figure 8.43 shows the system advising the user to reposition clamps when they are not aligned with the face and Figure 8.44 shows how the system responds when the positioning of the clamps are in accordance with the rules. Figure 8.45 shows the SIFD solution complete with hook clamps.

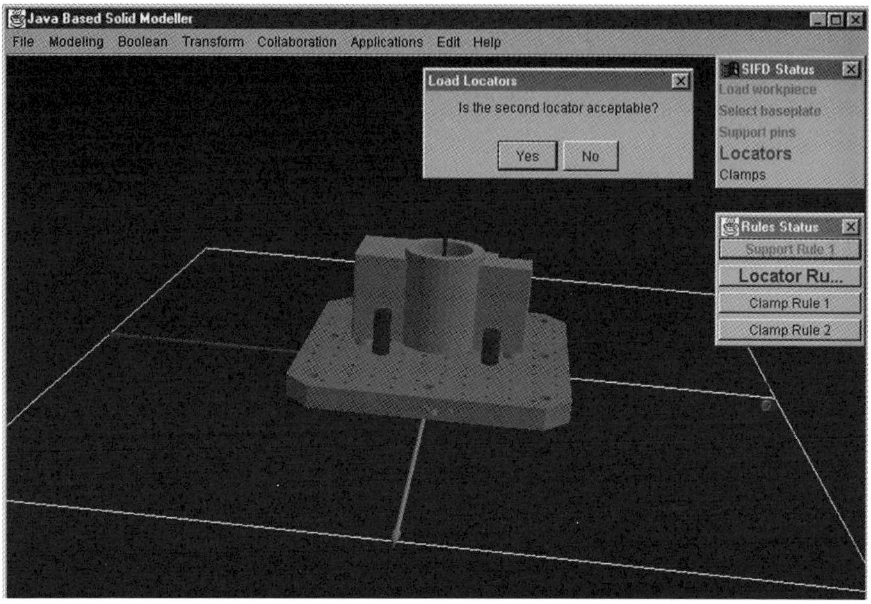

Figure 8.38 Selected locators loaded and positioned

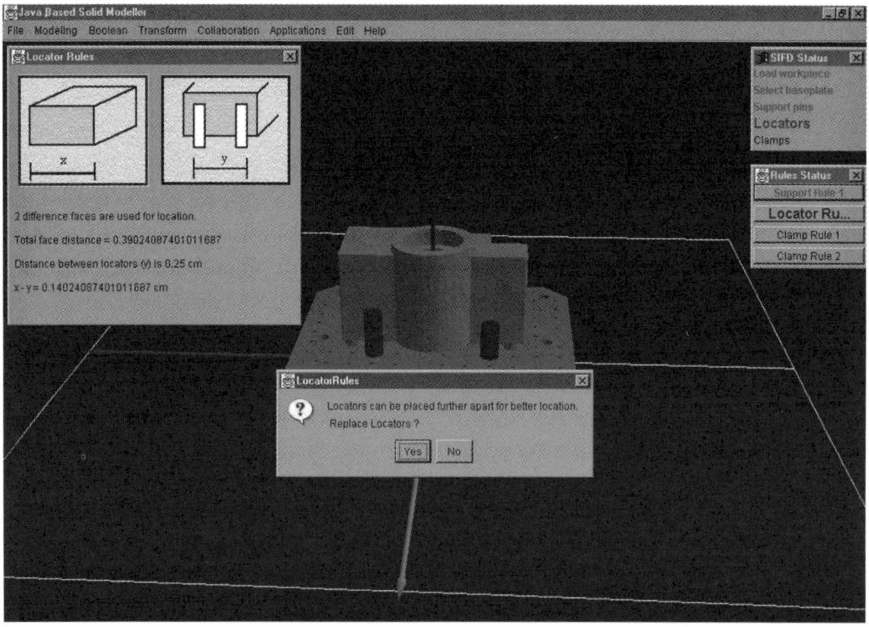

Figure 8.39 Locator rule analysis done on locators

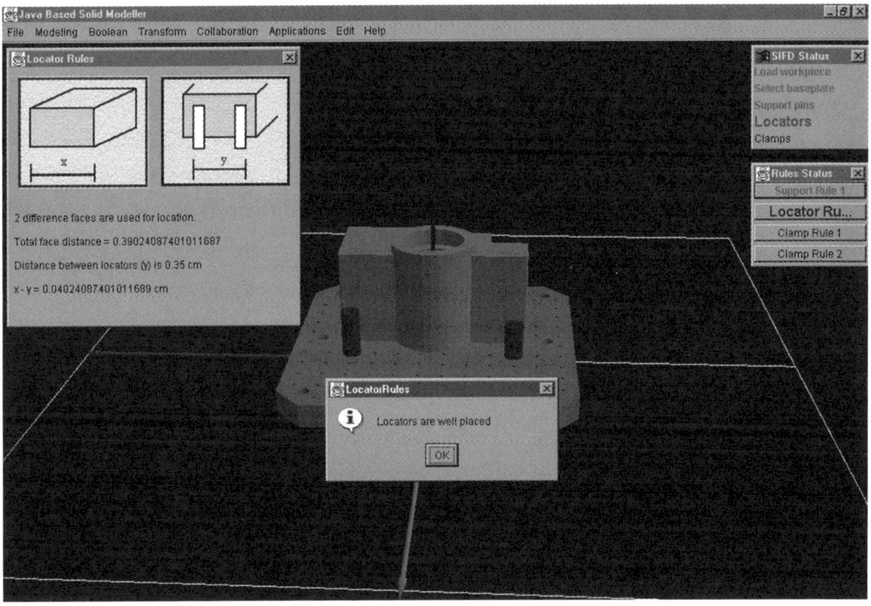

Figure 8.40 Analysis done on new locator positions

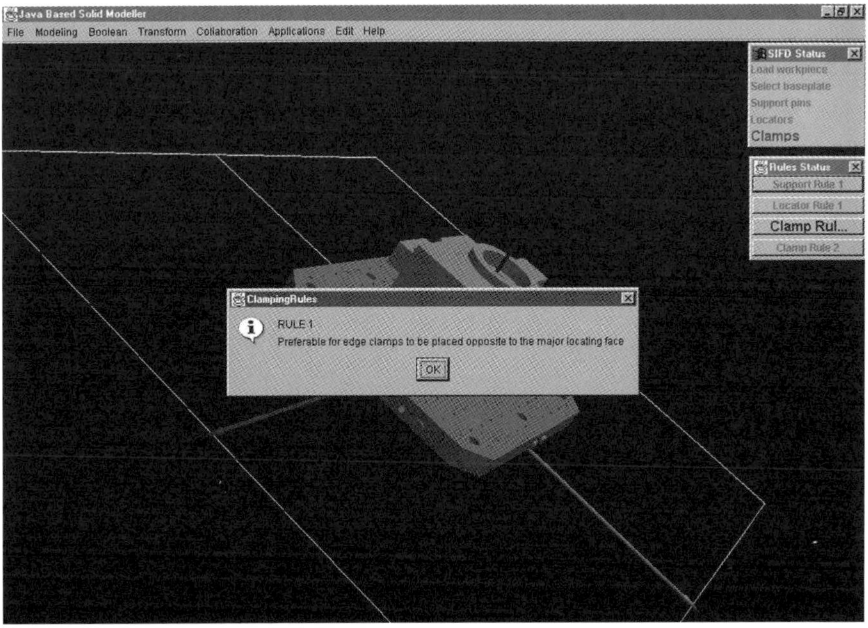

Figure 8.41 Clamp Rule 1 is invoked when the side clamp is selected

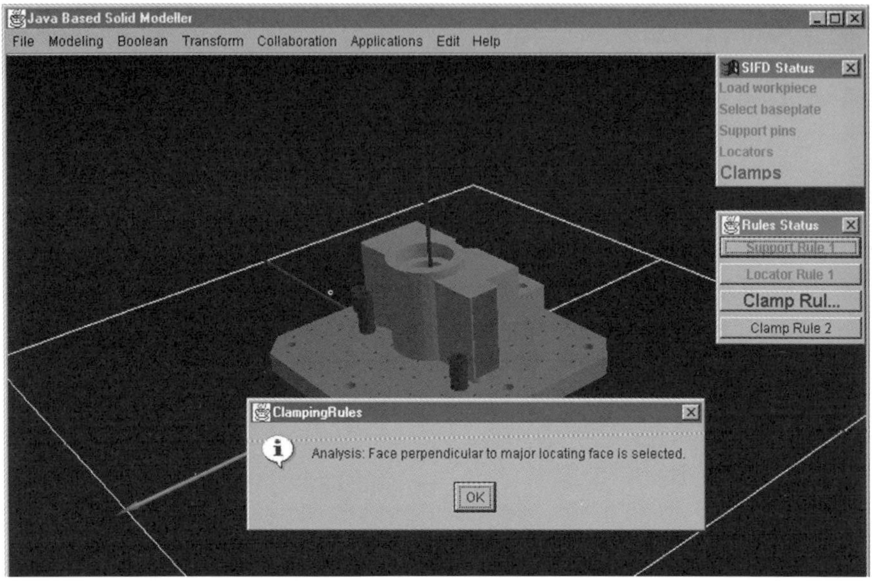

Figure 8.42 Analysis done on side clamp based on Clamp Rule 1

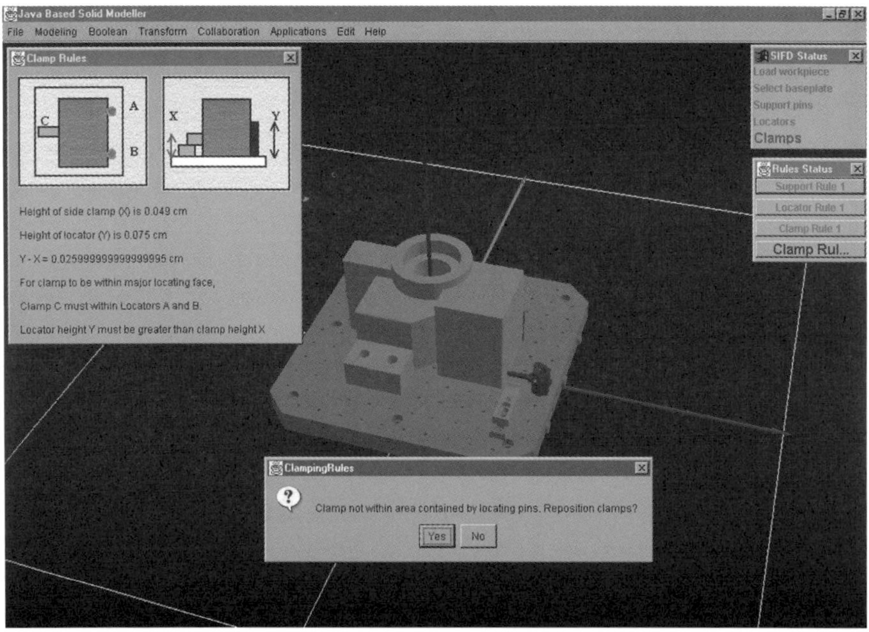

Figure 8.43 Analysis done on side clamp based on Clamp Rule 2

Figure 8.44 Analysis done on clamp at a new acceptable position

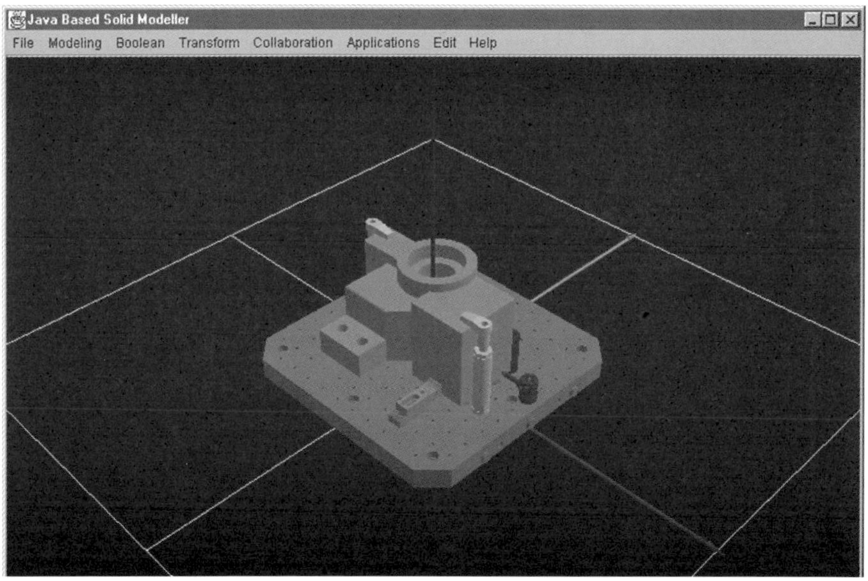

Figure 8.45 SIFD solution complete with hook clamps

8.6 Summary

An Internet-based fixture design has been developed and explained in detail. In terms of design capabilities, the SIFD system has overcome some of the main limitations of the existing dedicated fixture design systems. It has tackled the problem of data loss when the application is closed and also provided for other alternatives to the process of fixture designing, such as giving the user the ability to perform design changes on previous design setups. All these changes have an impact on achieving a truly collaborative design environment for fixturing.

Through the implementation of a few general fixture rules, the developed system has achieved the following: (1) obtaining an applicable fixture solution and (2) creating a user-friendly environment for fixture design. This will set the stage for expanding the market base of the fixture design application, as users who are not familiar with fixture design will also be able to use the system.

8.7 Bibliography

Arya, S. K. *Development of an Internet-Enabled Feature Based Modelling System.* B.Eng. Dissertation, National University of Singapore. 2001.

Balasubramanian, S. *Fixture-Based Design Similarity Measures for Variant Fixture Planning.* M.Sc. Dissertation, University of Maryland. 1999.

Dai, J.R., A.Y.C. Nee, J.Y.H. Fuh and A. Senthil kumar. An Approach to Automating Modular Fixture Design and Assembly. *Proc. Instn. Mechanical Engineer*, Part B, Vol. 211, pp.509-521. 1996.

Eary, D.F. and G.E. Johnson. *Process Engineering for Manufacturing*. Prentice Hall Inc. 1962.

Ferreira, P.M. and C.R. Liu. Generation of Workpiece Orientation for Machining using a Rule-based System. *Robotics and Computer-Integrated Manufacturing*, Vol. 4, No. 3-4, pp. 545-555. 1988.

Fuh, J.Y.H., C.H.Chang and M.A. Melkanoff. An Integrated Fixture Planning and Analysis System for Machining Processes. *Robotics and Computer-Integrated Manufacturing*, Vol.10, No.5, pp.339-353. 1993.

Kakish, J., P.L. Zhang and I. Zeid. Towards the Design and Development of a Knowledge-based Universal Modular Jigs and Fixtures System. *Journal of Intelligent Manufacturing*, Vol.11, No.8, pp.381-401. 2000.

Kiran, R.K. *Web Based CAD System.* M.Sc. Dissertation, National University of Singapore. 2001.

Kow, T.S. *Development of an Integrated, Automated, Interference-free Modular Fixture Design System.* M.Sc. Dissertation, National University of Singapore. 1999.

Mervyn, F., A. Senthil kumar, S. H. Bok and A. Y. C. Nee. Development of an Internet-enabled Interactive Fixture Design System. *Computer-Aided Design*, Vol.35, No.10, pp.945-957. 2003.

Senthil kumar, A. *A Knowledge-Based Generative Modular Fixture Design Methodology for Prismatic Workpieces*. PhD Dissertation, National University of Singapore. 1992.

Senthil kumar, A., V. Subramaniam and K.C. Seow. *Conceptual Design of Fixture Using Genetic Algorithms*. International Journal of Advanced Manufacturing Technology, Vol. 84, pp.79-84. 1999.

Shu, S.H. and J.L. Chen. A Modular Fixture Design System Based on Case-Based Reasoning. *International Journal of Advanced Manufacturing Technology*, Vol.10, pp.389-395. 1995.

Smith, C.S. and P.K. Wright. CyberCut: A World Wide Web Based Design to Fabrication Tool. *Journal of Manufacturing Systems*, Vol.15, No. 6, pp. 432-442. 1996.

Tan, B.C. *Development of An Internet-Enabled Interactive Fixture Design System*. B.Eng. Dissertation, National University of Singapore. 2000.

Wagner, R., G. Castanotto and K. Goldberg. FixtureNet: Interactive Computer Aided Design via the WWW. *International Journal for Human-Computer Studies*, 46, pp.773-788. 1997.

Yap, K.A. *Development of a Semi-automated Modular Fixture Design System*. B.Eng. Dissertation, National University of Singapore. 1999.

Index

List of Authors